HUMAN REPRODUCTION

L. M. Baggott
University of Exeter

CAMBRIDGE
UNIVERSITY PRESS

PUBLISHED BY THE PRESS SYNDICATE OF THE UNIVERSITY OF CAMBRIDGE
The Pitt Building, Trumpington Street, Cambridge CB2 1RP, United Kingdom

CAMBRIDGE UNIVERSITY PRESS
The Edinburgh Building, Cambridge CB2 2RU, United Kingdom
40 West 20th Street, New York. NY 10011–4211, USA
10 Stamford Road, Oakleigh, Melbourne 3166, Australia

First published 1997

Design by Hart McLeod, Cambridge

Printed in the United Kingdom at the University Press, Cambridge

Typeset in Palatino 9.5pt

A catalogue record for this book is available from the British Library

ISBN 0 521 46914 7 paperback

Cover picture: Science Photo Library

Acknowledgment
I acknowledge with thanks the contribution of friends and colleagues from the Department of Obstetrics and Gynaecology at the Royal Devon and Exeter Hospital (Heavitree) who gave me the opportunity to write this book, and also that of Professor Ted Wragg of the University of Exeter for his encouragement. I would like to dedicate it to my undergraduate students, past and present, for their unfailing enthusiasm and interest in this subject.

Contents

ONE

Human reproductive strategies

It has been said that the final aim of all human love relationships between men and women is more important than all other ends in life, because what it determines is nothing less than the composition of the next generation. It is perhaps no coincidence, then, that reproduction – involving, as it does, sexual intercourse – has come to occupy such an important part of our lives. It is the driving force of evolution, and hence generates the great variety of living organisms which inhabit, and have inhabited, the Earth, including, of course, ourselves. However, notwithstanding the power of evolution, or perhaps because of the force of it, sex has come to mean much more to us than reproduction.

Normal sexual intercourse has been defined as anything erotic which gives pleasure to both partners, who are consenting adults, and does not hurt anyone. This clearly holds good for a wide variety of sexual functions within loving relationships. In this context, the word love should be taken to mean the desire for the good of the other person; wishing for gain, or to exercise power over another person does not occur in a loving relationship.

Clearly, there are not only emotional but also physical risks associated with having a sexual relationship. Unwanted pregnancies and sexually transmitted diseases can all too easily follow from unprotected sexual intercourse (see chapter 6 for more information about contraception). However, though it bears great personal responsibility and requires a degree of self discipline, sexual activity is normal and healthy and can be a joyous part of life.

Human beings are unique in the living world in that although we share with all other organisms an overwhelming drive to procreate, we are the only species consciously to limit the number of our offspring and to outlive, often by many years, our reproductive capacity. Like all other biologically based aspects of the human condition, the complex nature of human reproduction has come about in response to the various selection pressures to which it has been subjected. This is an underlying theme of this book, so it is appropriate, therefore, that we begin our study of human reproduction with a brief consideration of its evolution.

1.1 An evolutionary perspective

Physiological adaptations

In evolutionary terms, reproduction is the most important defining characteristic of life (see also chapter 9). However, although it is vital for the survival of the species as a whole, it is not essential to the individual. Amongst living things there is an almost infinite variety of strategies for ensuring reproductive success, and humans are no exception to this. Human reproduction, though, is unique in two important respects: the extended period of development before sexual maturity is reached, and the lack of a defined mating season, or 'heat'.

In non-primate mammals, there is no **prepubertal** growth spurt: rodents, for example, become sexually mature whilst they are still growing. Primates, including humans of course, have a developmental mechanism which prevents the maturation of their **gonads** (ovaries and testes) until after they have undergone an increase in growth rate before adolescence. This provides an extended period for learning. Because they are small and relatively weak, the young can be more easily cared for, and they can learn through play without injuring each other too much. At **puberty** they quickly become much larger and stronger just as they mature sexually, so as they become adults they are capable of caring for their own smaller offspring.

In mature men the testes continually release **sperm** which is stored in the long, coiled duct of the epididymis. From time to time this is released via the vas deferens and mixed with secretions from the seminal vesicle and prostate gland to form seminal fluid, which is ejaculated from the penis during orgasm. Although there is no evidence that the central nervous system is made aware of the need for emptying of the epididymis or accessory glands, there is no doubt that the individual is fully cognisant of the special satisfaction of this event. In women, production of **oocytes** (ova or eggs) from the ovaries is cyclic, and involves no comparable accumulation of large numbers of gametes or quantities of fluid. However, the orgasm experienced by many woman is a sexual satisfaction comparable with that of male ejaculation, and has contributed to the heightened sexuality which is a particularly human characteristic. A detailed description of the functioning of the mature male and female reproductive tracts appears in chapters 2–5.

Humans are physiologically adapted for a frequent and repeated need for sexual stimulation and gratification. That sexual intercourse can take place at any time is an important factor in human social evolution. In contrast, mating episodes usually occur only around the time of ovulation in most other primates. In many of these species, such as the baboon and chimpanzee, changes indicating her receptivity to the male occur in the female external genitalia and behaviour. Human women, on the other hand, rarely experience any external sign that they are ovulating: occasionally there may be slight bleeding in the middle of the menstrual cycle, and although there is a slight rise in body temperature at the time of ovulation, the woman is rarely aware of it (chapter 7). There is no evidence that intercourse occurs more often at this time during the woman's cycle.

As **gametes**, sperm outnumber eggs by several orders of magnitude around the time of fertilisation. This means that sperm, from either one or several acts of intercourse with one or more partners, must compete with one another to fertilise the egg. This is a subject which has long been a source of fascination for reproductive and evolutionary biologists. Some suggest that this **sperm competition** may form the physical basis of much sexual behaviour between men and women. On a purely biological level, sperm competition takes place within each of the various strategies for human procreation: monogamy (having only one sexual partner at a time); serial monogamy (having more than one partner, but only one at a time); polygamy (having more than one female – or less usually male – partner at a time); and promiscuity (having several sexual partners at a time). Although it is the norm in Western society, not all human societies are monogamous, and in some the man takes little or no responsibility for raising children. This brings us to the question of whether the various observable physiological adaptations for reproduction may be linked to the human sexual behaviour which forms such an important part of our social life.

Social and behavioural adaptations

Just as anatomical and physiological adaptations are responses to evolutionary selection pressures, so behavioural patterns have evolved to help to ensure the reproductive success of our species. The study of the biological basis of social behaviour in animals, including humans, is called **sociobiology**, and it has helped to explain how the evolutionary pressures which have brought about all other survival adaptations have also played their part in the development of behaviour, including subtle examples related to reproduction, such as altruism, parent–offspring relationships, nepotism and sexism.

Although little detail is known, it is certain that fertility was a prevailing concern of people at the dawn of human evolution. Human courtship displays and reproductive strategies are the most complex and elaborate in the animal kingdom, and there seems little doubt that these patterns are based upon those of our early ancestors. While little is known of these, study of related species suggests that communication between the females and males which drew them together and promoted mating, and the reproductive strategy which ensured the efficiency of mating, were of fundamental importance.

A characteristic of human sexual and mating behaviour is the formation of relatively long-lasting pair bonds. There is evidence that this also occurs in gibbons, but is not otherwise generally seen in primates. It is thought that during Pleistocene times, humans started to hunt in an organised way, and the men who went hunting probably spent extended periods away from the rest of the group. It is therefore possible that at this time pair bonds evolved for the protection of the mothers and their offspring, probably in response to the need to nurture the offspring for many years to ensure their survival to reproductive maturity. Monogamy, then, may have arisen as a strategy by which the woman obtains the permanent attentions of a man who will help her fend for their offspring, and the man is rewarded by having his sexual drive gratified by a woman who is sexually receptive throughout her reproductive cycle. Also, and most importantly, monogamy helps to ensure that the genes of each partner are passed on to the next generation. Sexual pleasure has therefore become very important in our species. If women could only enjoy sexual satisfaction at the time of ovulation (2–3 days every 28 days) this might jeopardise their ability to form lasting pair bonds. This is an example of structural and functional adaptations forming the basis of the social development and change which ultimately becomes an evolutionary selection pressure.

It is not known exactly when the family became a feature of human society, but it seems reasonable to speculate that it is a survival strategy based on reproductive pressures. From very early times, societies have imposed rules which govern the biological and social consequences of human sex drive to produce stability in communities. The sexual drive in humans is thus one of the primary factors which have contributed to the development of civilisation, and the structure of today's society is founded upon it. In many cases, particularly in the Western World, this has resulted in monogamy being the norm.

However, for our species, copulation has a greater significance than just reproduction. Inevitably, human sex drive also has less favourable influences on behaviour: it can be used in assault, for wielding power in relationships and for material gain. This is so because of the complexity of human behaviour, which is made possible by the relative size of the human cerebral cortex. A large brain is probably the single most distinguishing human feature. The evolution of the brain has enabled the development of an infinite variety of human behaviour, and a great deal of this behaviour influences and is influenced by sexuality.

1.2 Ethnic and religious factors

As early people progressed from their lives as hunter–gatherers towards the first attempts at agriculture, the drive to produce their own offspring was translated to the struggle to ensure the fruitfulness of their livestock and crops. Promotion of fertility was central to the religious beliefs of these people, and fragments of these beliefs survive in the folk memory to the present day: Easter eggs, May Day ceremonies and spring weddings, to name but a few. At the heart of most modern religious systems is the belief that the fundamental difference between humans and other animals is that we have free will. One manifestation of this, as far as reproduction is concerned, is that we have broken away from biological necessity – each culture and subculture has developed its own patterns of behaviour.

The subject of ethnic and religious views on sexuality is vast and fascinating, but well beyond the scope of this book. Chapter 9 deals with some of the moral and ethical aspects of this which are deeply rooted in our various belief systems. Study and discussion of the many ethnic approaches and religious beliefs about sexuality and reproduction may help us to come to a clearer and more tolerant understanding of one another in our increasingly pluralist society.

1.3 Human reproduction – an overview

Issues surrounding human reproduction have been central to the development of society, because they are of such fundamental importance to us all. The scientific revolution which has enabled us to manipulate reproduction to the extent that we can control fertility and, to some extent, infertility, has gathered enormous momentum in the past decade, and this has raised public awareness of these issues. However, perhaps because it is so central to our lives, the subject of human reproduction is extremely complex, and coming to an understanding of it requires consideration not only of sociology and psychology, but also of the underlying biochemistry, as well as the structure and function of the reproductive cells, tissues and organs.

While focusing on the biology of human fertility and infertility, this book will attempt to reflect the complexity of this topic by, where appropriate, setting the biological issues into a social context.

This is because our lives are, to a very great extent, controlled by others, and the pressure produced by the exponential increase in human population has been an important factor in the rise of the technology which has enabled us to control it.

TWO

Gamete production

Omnis cellula e cellula, 'All cells from cells' stated the German clinician Rudolf Virchow in 1855. From the beginning of evolution, the cells constituting the bodies of all life on Earth have come from pre-existing cells. More than any other characteristic of life, the ability to reproduce distinguishes the living from the non-living, and the process of reproduction is based on the cell cycle, the turn of which is dependent on cell division.

2.1 Cell cycle

Along with all other living things, except bacteria and cyanobacteria, the bodies of human beings are made up of eukaryotic cells. The vast majority of these cells are continually renewed and replaced by a form of cell division. This complex but ordered process involves duplication of the genetic material (DNA) in the nucleus of the cell, followed by its division into two equal portions, the migration of the portions to opposite sides of the cell (this part of the process is called **mitosis**), and finally the division and partition of the cytoplasm and its contents (**cytokinesis**).

A full cell cycle is that period from the end of one mitosis to the beginning of the next (figure 2.1). Division is the shortest phase of the cell cycle, and represents 5–10% of the total cell cycle time. Known as the M (or mitotic) phase, it is separated from the next M phase in the cycle by **interphase**, in which the cell is metabolically active and performs whatever functions for which it is adapted. After M phase is G1, or a gap phase, which occupies about 30–40% of the cycle during which the cell is metabolically active and is often increasing in size following division. In genetic terms it functions with the same amount of DNA as it received from the parent cell. During the S phase (30–50%) which follows G1, the cell synthesises new DNA. Another gap phase, G2, follows (10–20%) during which the cell prepares for the next mitosis, and another cycle, by replicating its organelles. So, for every cell that enters mitosis two emerge, each continuing through the subsequent cycle as independent entities.

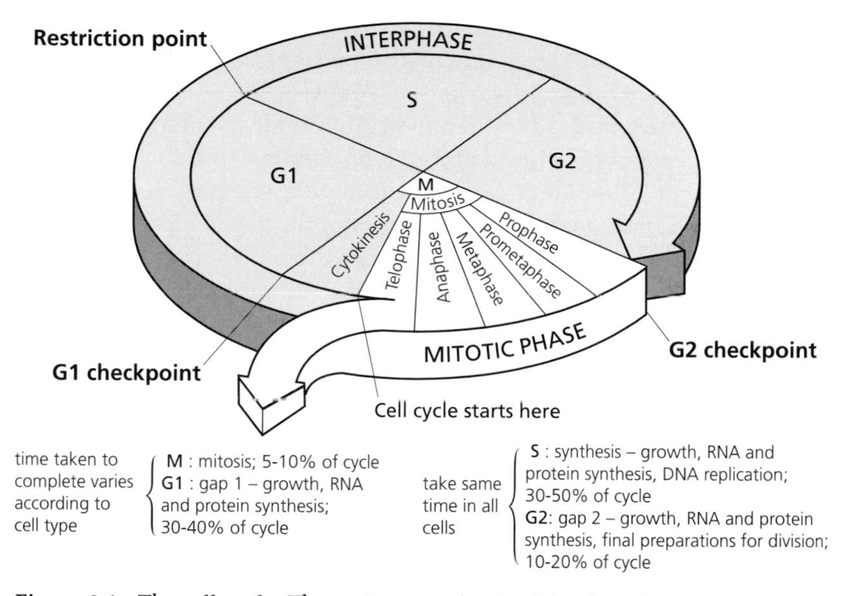

Figure 2.1 The cell cycle. The entire reproductive life of a eukaryotic cell is a 4-stage cycle. A full cycle is that period from the end of one mitosis to the beginning of the next. M is the mitotic phase during which the nucleus divides. This is followed by G1, the first gap phase, after which comes the S phase when the cell synthesises new DNA. A second gap phase, G2, follows, and at the end of this the cell enters M once more.

The length of a cell cycle varies considerably from one cell type to another, and from one species to another. In humans there are about 10^{13} cells, each different type cycling at a different rate, the shortest cycles being those which take place during early embryonic development. In the liver cell, however, the cell division cycle can take more than a year. Two notable exceptions to this are nerve and muscle cells, which do not divide but remain arrested in the G1 phase for their entire life.

Control of the cell cycle is an important focus of current research in cell biology, because of its relevance to our understanding of cancer. In all cells the stages S to M always take a fixed length of time; G1 is the phase which varies. At the end of G1 is a point of no return, known as the **restriction point**. It is thought that a protein accumulates (or not) during G1 which enables the cell to get past the restriction point and enter the S phase. Once the restriction point is passed, the cell cycle turns through S, G2, M and G1 again. Other such checkpoints on the cell cycle are thought to operate at the end of G2 and at metaphase.

Duplication of the DNA is made more reliable by the grouping and folding of the molecules into compact structures called **chromosomes**. The chromosomes are very long molecules, short sections of which encode for specific proteins. These sections are called **genes**, and the entire genetic complement of a cell is known as the **genome**. The fact that the process of

DNA replication, and hence duplication of the entire genome, so rarely goes wrong, and faithful copies are made time after time, is the molecular basis of the amazing miracle of life.

All cells are genetically programmed to divide a certain number of times, and this number varies with cell type. One notable outcome of the Human Genome Mapping Project, a world-wide scientific collaboration to identify all human genes, is the discovery of the genetic mechanism for this 'programmed cell death'. It is not impossible that scientists may, in the not-too-distant future, be able to manipulate this gene such that cell death might be postponed, and life possibly prolonged. Inevitably, this will raise a huge moral debate.

2.2 The significance of meiosis

Multicellular organisms reproduce sexually. This process is the result of the fusion of two cells, as opposed to the fission of one cell to make two as in the case of unicellular asexual reproduction. The significance of this is that the resulting individual possesses a new and unique genome which is quite distinct from that of either parent. It is in the obvious and subtle differences between parent and offspring that the potential for change through variation lies. The process by which these differences are produced involves a type of cell division which occurs during the production of the gametes (sperm and oocytes, or eggs) and which results in a halving of the genome. Every eukaryotic organism has a characteristic **diploid** chromosome number – in humans it is 46. This means that in every **somatic** cell (all body cells except the gametes) there are 46 chromosomes. The type of cell division which produces the gametes is known as **meiosis**, and it results in each human sperm and oocyte containing 23 chromosomes – the **haploid** number. During the process of fertilisation the diploid number of chromosomes is restored, and every subsequent turn of the cell cycle during embryological development, growth to maturity and body maintenance results in two diploid daughter cells. However, very early in embryonic life (chapter 4) a germ cell line is determined, and two distinct cell populations result: that is somatic cells and germ cells. The germ cells ultimately give rise to the gametes. These two cell lines remain distinct throughout life, and the processes of mitosis and cytokinesis retain the full genome in all the cells of the body except the germ cells.

A full account of the process of meiosis can be read in *Heredity and human diversity* by Stephen Tomkins (CUP, 1989), but the essential point about it is that it consists of two cell cycles, only the first of which involves duplication of the DNA. The prophase of the first meiotic division is specialised to allow exchange of sections of genetic material between the chromosomes. This process is called crossing-over, and results in a recombination of genes which greatly increases the variability of the gametes. After the extended and complex prophase, the chromosomes line up on the equator of the cell in homologous pairs so that the individual members of each pair are pulled to opposite poles. Separation and migration of the chromosomes is followed by a short interphase which separates the two divisions, during which there is no DNA synthesis in the S phase. The second meiotic division is quicker and simpler than the first, and the events are very similar to those in mitosis: the chromosomes line up on the equator of the cell, the centromere of each chromosome divides and the sister chromatids separate.

2.3 Hormonal control of gamete production

The human reproductive system, in common with that of all other mammals, is under the control of the central nervous system (CNS). The gonads (ovaries and testes) have two main functions: to produce gametes and to produce hormones. The male sex hormones are called **androgens**, the most important of which is **testosterone**, and the female hormones are the **oestrogens** and **progesterone**. Production of the sex hormones by the testis and ovary is under the control of gonadotrophic hormones secreted by the anterior lobe of the pituitary gland just beneath the brain (figure 2.2). The gonadotrophic hormones are f**ollicle stimulating hormone** (FSH) and **luteinising hormone** (LH). In common with all endocrine secretions, they are released directly into the bloodstream in which they travel until they reach their target organs. These organs are recognised by specific receptor proteins in the cell surface membranes of the target organ cells.

The pituitary gland is itself under the control of a particular part of the floor of the brain, called the **hypothalamus**, to which it is physically connected by a specialised stalk containing portal blood vessels. These vessels provide a route for transfer of information from the hypothalamus to the pituitary gland. The hypothalamus contains neurones which, unlike other neurones, secrete hormones instead of **neurotransmitters** that help in the transmission of nerve impulses. These hormone-secreting neurones are called **neuroendocrine** cells, but in many other respects they function similarly to ordinary neurones: they have cell bodies, axons, dendrites and so on, and they receive neural stimulation from other neurones via synapses. The hormones secreted by the hypothalamus, sometimes called **releasing factors**, enter the portal system in the pituitary stalk and, on reaching the anterior lobe of the gland, stimulate it to release its hormones. The hypothalamic releasing factor responsible for secretion of the gonadotrophic hormones FSH and LH is called **gonadotrophin releasing hormone (GnRH).**

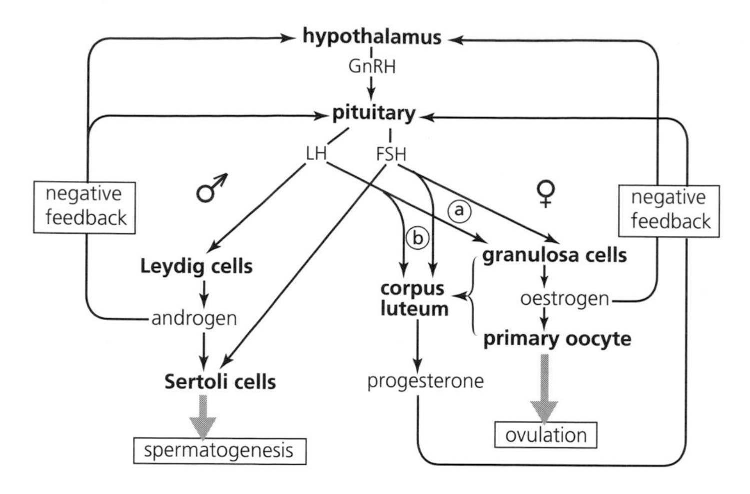

Figure 2.2 Hormone interactions between the hypothalamus, pituitary and gonads (*a*) follicular phase of the ovarian cycle; (*b*) luteal phase of the ovarian cycle.

The significant point about the involvement of the hypothalamus is that it is here that neural stimuli are changed into chemical stimuli. In some animals this is extremely important, as successful reproduction is often linked to a number of external factors, such as the seasons, proximity of others of the species, and so on. The animal is made aware of these factors by a variety of visual, tactile and olfactory cues, which are fed into the CNS and eventually reach the hypothalamus where they are processed, amplified and changed into chemical information in the form of GnRH. The information is further amplified and passed on by the pituitary to the gonads in the form of FSH and LH. The gonads respond by producing sex hormones which act on a variety of target organs, including the pituitary itself.

Although the extent to which external factors affect human fertility is not fully understood, this integrated CNS–endocrine information system coordinates all aspects of reproduction, including sexual function and maintenance of secondary sexual characteristics.

2.4 Gametogenesis

The cells which give rise to the gametes (eggs and sperm) are known as **primordial germ cells** which arise in the embryo during the fourth week after fertilisation. During the fourth to sixth week of embryonic development these, as yet unspecialised, cells migrate to a position within the embryo where they will ultimately bring about the formation of the gonads. During this migration they multiply continuously by mitosis. A fuller account of the development of the reproductive system can be found in chapter 4. The embryonic germ cells develop as **stem cells**, and here, nourished and protected by somatic cells of the gonad close by, the stem cells multiply by mitosis. This is known as the phase of proliferation. In the female, the supporting cells in the ovary are called **follicle cells**, and in the male the supporting cells in the testis are called **Sertoli cells**. The germ cells themselves are known as **gonia**: hence **oogonia** and **spermatogonia**. The production of mature gametes from these gonial cells is called respectively **oogenesis** and **spermatogenesis**, and collectively **gametogenesis**. Although the actual timing is different in males and females, gametogenesis in both genders consists of two further distinct phases: growth and maturation, during which meiosis is completed. Gametogenesis is under direct hormonal control, and involves considerable differentiation of the cell structure and reduction to half of the chromosome number (figure 2.3).

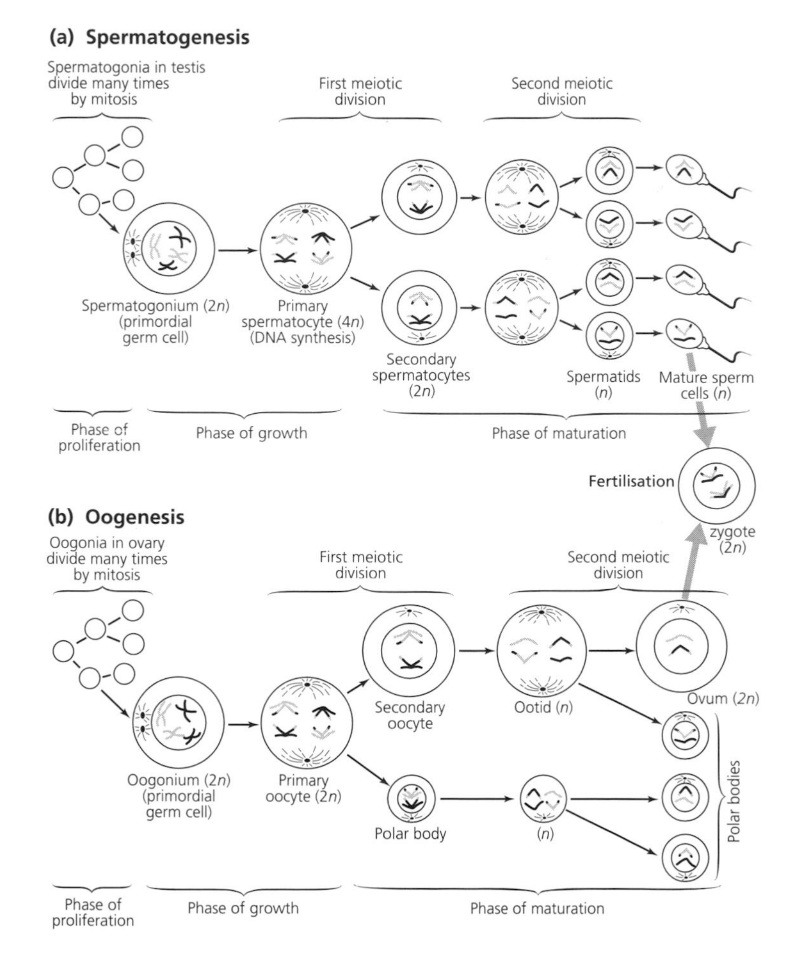

Figure 2.3 Chromosomal changes during gametogenesis.

2.5 Spermatogenesis

Mammalian sperm are highly specialised cells. Amongst the various species there is a staggering variety of shapes, motility patterns and functional strategies. The sperm cell has only one purpose, to fertilise, and all the adaptations of structure and function which have come through the course of evolution have occurred in response to this overriding aim. The testes have also become adapted as highly efficient production lines of vast numbers of these extraordinary cells (figure 2.4).

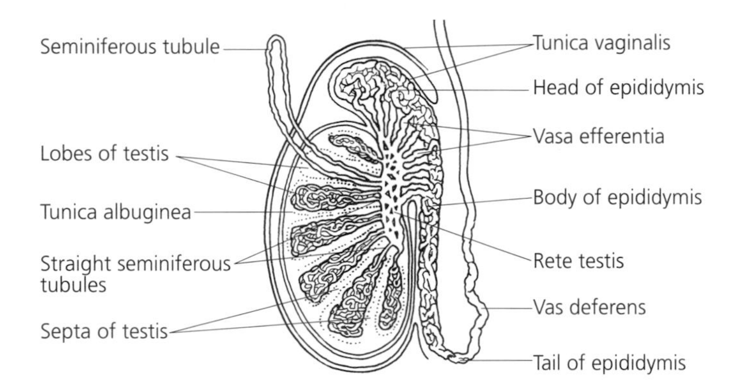

Figure 2.4 Section through human testis.

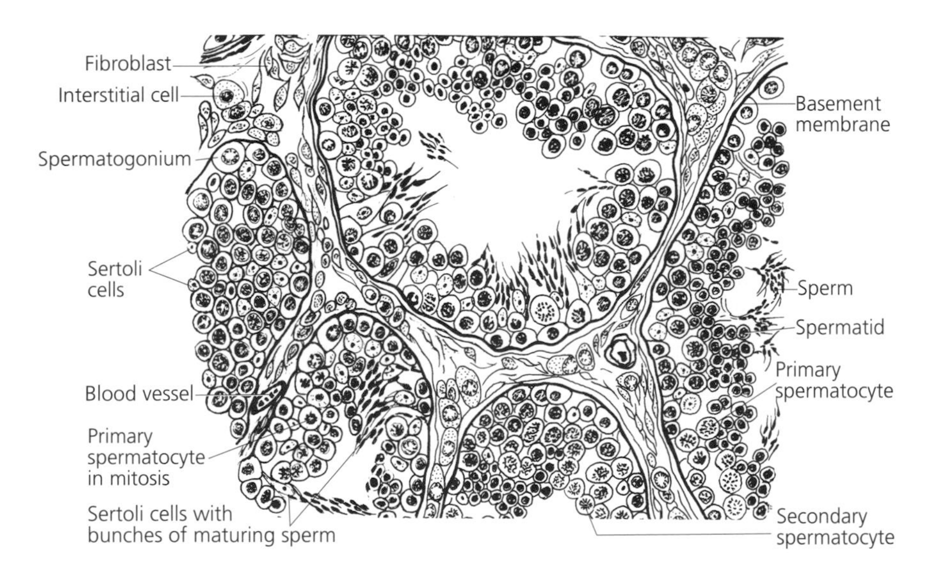

Figure 2.5 Section of human testis (x470).

The testis consists of two separate compartments: one which produces the sperm (which in turn transmit the genes to the embryo after fertilisation), and one which produces the androgens upon which male sexual and reproductive function depend. Sperm develop within the seminiferous tubules, while androgens are synthesised between them, in specialised interstitial cells called **Leydig cells** (figure 2.5). The two testicular compartments are separate physiologically, and this means that the microenvironment in each is quite different.

The seminiferous tubules are lined with **Sertoli cells** (figure 2.6) which are columnar in shape with narrowing tips extending into the lumen of the tubule. These cells are connected by gap junctions, through which they communicate. The developing germ cells remain in very close association with the Sertoli cells throughout the process of spermatogenesis. The spermatogonia are found at the base of the Sertoli cells near to the basal membrane and, as they progress through the stages of spermatogenesis, the germ cells move towards the lumen of the seminiferous tubule. Spermatozoa are released into the lumen from the tips of the Sertoli cells.

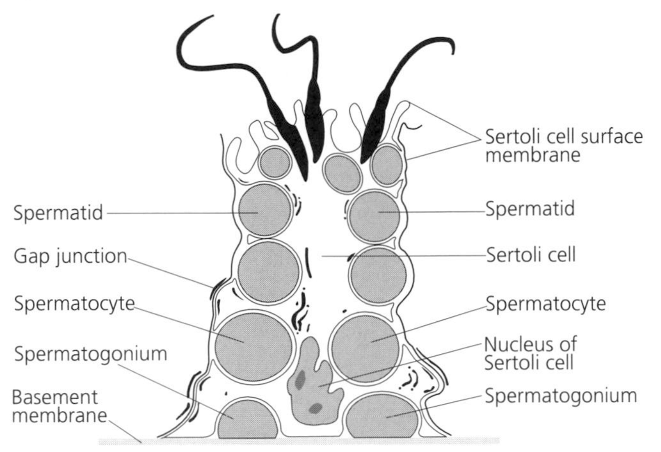

Figure 2.6 Sertoli cells and spermatogenesis. The Sertoli cell is intimately involved with the developing male gametes during spermatogenesis. As the process of spermatogenesis unfolds, the developing gametes move progressively towards the apex of the cell, away from the basement membrane (figure 2.5). Fully formed spermatozoa are released into the seminiferous tubule.

However, the sperm are not yet mature. They have to spend a further period in the epididymis (figure 2.4) during which they become motile, and various changes to the cell surface membrane occur. Epididymal sperm maturation is discussed in chapter 3.

2.6 Oogenesis

Although apparently less specialised than sperm, differentiation of the oocyte is more complex in terms of contribution to the succeeding generation, because it contains not just the haploid genome, but also very important cytoplasmic components on which early embryological development depends.

Another important difference between oogenesis and spermatogenesis is that the number of primordial germ cells which develop into oogonia in the female fetus is finite. By the time the fetus is five months old, the mitotic or proliferative phase of gametogenesis will have resulted in about 7 million oogonia in the developing ovaries. At birth this number will have fallen to perhaps a million or less. Just before birth the proliferative phase terminates, and all oogonia enter the first meiotic division to become **primary oocytes** (figure 2.3). This means that at this point in her life, the baby girl has all the potential oocytes she will ever have and, if she subsequently loses these, they cannot be replaced from stem cells, as would be possible in the testis.

During the first meiotic prophase, the oogonia become surrounded by ovarian stroma cells which secrete basement membranes, and **primary follicles**, containing primary oocytes, are formed. These events occur during the early stages of the first meiotic prophase by the end of which further development of the primary oocyte is arrested. The chromosomes are still surrounded by nuclear membranes within nuclei known as **germinal vesicles**: the primary oocytes are said to be at the **dictyate** stage. They all remain in this state until puberty, when a small number of primary follicles is regularly recruited for further development. Some primary follicles may not undergo further development for perhaps another 50 years, but they remain in this suspended state until the signal arrives to continue the first meiotic division.

Pre-antral follicular development

In fact, the next stage of follicular development involves growth of the oocyte and follicle. There is also considerable protein and RNA synthesis, but meiosis is not immediately resumed. The **granulosa cells**, which surround the oocyte, become several layers thick and secrete a glycoprotein which forms a continuous layer overlying the oocyte cell surface membrane. This layer is known as the **zona pellucida**. The stroma cells of the ovary become enriched with blood vessels and form a loose layer, called the **theca** which surrounds the granulosa layer. This first phase of follicle development, the pre-antral phase (figure 2.7), involves growth and increasing complexity of the structure of the follicle. This phase is not dependent on hormonal stimulation but, towards the end of the pre-antral phase, the granulosa cells develop membrane receptors for oestrogen and FSH and the theca cells develop receptors for LH. Gonadotrophic hormone stimulation (FSH and LH) is then directly responsible for the next, or antral, phase of follicular development.

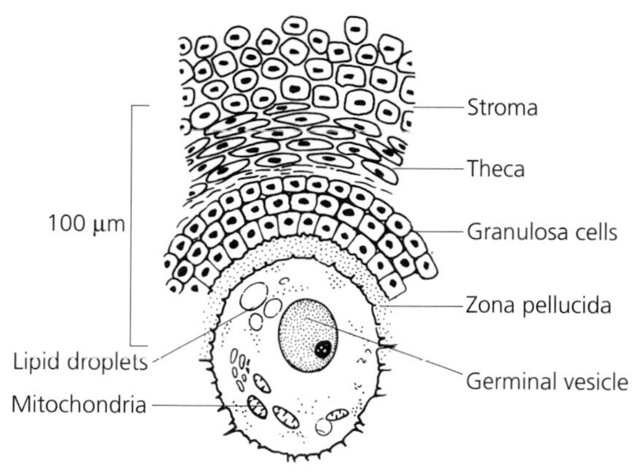

Figure 2.7 Pre-antral phase of follicular development in the ovary. During this time the primordial follicle, containing the dictyate oocyte, begins to grow from a diameter of about 20 μm to around 300 μm.

Development of the Graafian follicle

Antral, or **Graafian, follicles** (figure 2.8) develop as a result of proliferation of cells in the granulosa and theca, and accumulation of fluid between the granulosa cells. This follicular fluid eventually gathers together to form the fluid-filled **antrum**. The oocyte, which is still in the dictyate state, is surrounded by a mass of granulosa cells known as the **cumulus oophorus**. This is attached to the rest of the granulosa layer by a thread of cells. Throughout the antral phase of development, the follicles produce an increasing output of steroid hormones, mainly oestrogens, and just before ovulation there is a rise in the level of oestrogen in the blood.

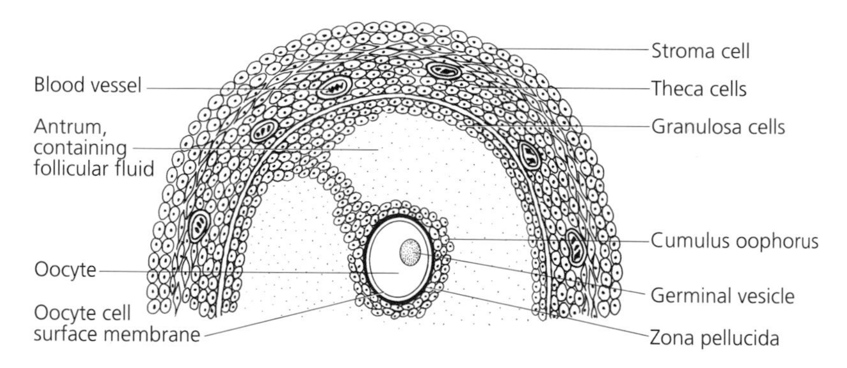

Figure 2.8 Structure of the mature Graafian follicle.

Preovulatory follicular development

The third stage of follicular development is the preovulatory phase, during which the oocyte matures and meiosis is resumed. It ends at ovulation, when the oocyte is released. The preovulatory phase is entirely dependent upon LH. The development of receptors for LH on the follicular cells coincides with a surge of LH from the pituitary gland. When this LH surge occurs, the nuclear membrane of the oocyte breaks down, and meiosis is resumed, ending the dictyate stage. The remainder of the first meiotic division proceeds but ends in an unequal division of the cytoplasm. This produces a large secondary oocyte and a smaller, extruded, first polar body, which contains half the chromosomes but only a very small amount of the cytoplasm. During metaphase, the chromosomes in the secondary oocyte line up along the equator of the cell but then, for a reason not clearly understood, arrest for the second time. It is at this stage of the second meiotic prophase that the oocyte is ovulated (figures 2.9 and 2.10). The cytoplasm undergoes further changes at this time. Synthesis of lysosome-like vesicles by the Golgi body occurs, and these organelles migrate towards the cell surface membrane. Known as **cortical granules**, they play a role during sperm–oocyte binding. New proteins are also produced which further prepare the oocyte for the events of fertilisation.

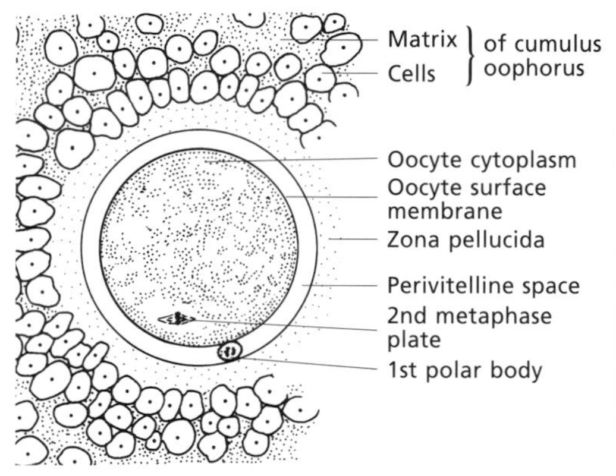

Figure 2.9 Structure of the newly ovulated oocyte.

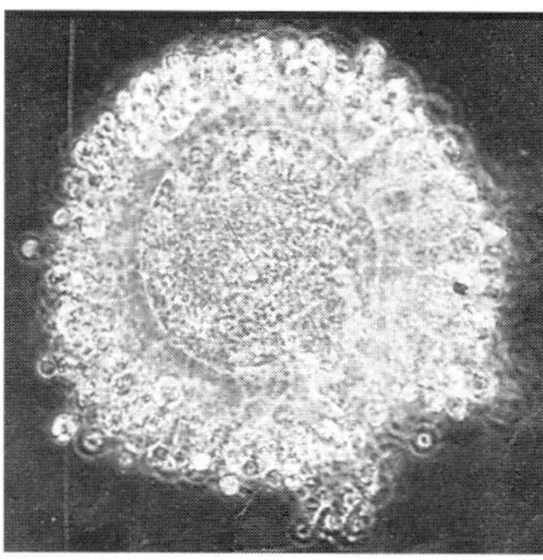

Figure 2.10 Electron micrograph of ovulated oocyte.

2.7 Ovulation

As ovulation approaches, the volume of fluid in the follicle increases, and the rise in pressure thins the wall of granulosa cells. The theca cells regress because of the surge in LH, and the mature Graafian follicle bulges from the wall of the ovary. The oocyte is still attached to the weakening surface of the follicle by the thin thread of granulosa cells. The epithelial cells covering the follicle lose blood vessels at a particularly weak point, and then appear to degenerate. This causes the follicle to rupture so that the fluid in the follicle and the oocyte, surrounded by its mass of cumulus cells, are expelled directly into the peritoneal cavity. The oocyte in the fluid is guided into the mouth of the Fallopian tube by **fimbriae**, fingerlike membranous projections at the opening of the tube. Ovulation takes place within about 12 hours of the end of the dictyate stage and the resumption of meiosis.

Multiple ovulation can occur if more than one follicle matures simultaneously. This may lead to more than one oocyte being fertilised, subsequently implanting and developing. In fact it is thought that this is not an uncommon event, as ultrasound examinations show that potential twins are often present early on in a pregnancy which results in a single live birth. There is a considerable danger of multiple ovulation when ovarian stimulation is carried out, for example during fertility treatment (see chapter 8). However, under natural circumstances, about 80 in 10 000 pregnancies result in twins born from two separate oocytes being fertilised (**dizygotic** twins). The incidence of dizygotic twins is influenced by genotype of the mother and possibly by environmental factors. Higher order births as a result of multiple ovulations occurring naturally are extremely rare.

2.8 Atresia

If a follicle is not sufficiently developed when the hormonal levels reach the critical point, it will undergo the process of **atresia**. An atretic follicle has an outer layer of granulosa cells which contain many lipid droplets, show reduced protein synthesis activity and have small irregular nuclei. The oocyte dies, and the follicle is subsequently broken down by the action of white blood cells, eventually remaining only as scar tissue in the stroma of the ovary.

2.9 Formation of the corpus luteum

The empty follicle collapses, and a clot forms within it containing some blood cells and remaining granulosa cells. These become rich in mitochondria, SER, lipid droplets and a carotenoid pigment known as **lutein** which gives the corpus luteum (CL) its yellowish colour. The whole structure begins to secrete progestogens, but is dependent on LH for its continuing function. If conception does not occur, the CL begins to degenerate after fourteen days, the lutein cells dying, and the output of progesterone falls. This process is called **luteolysis**. A whitish area, the **corpus albicans,** appears in its place, and this is eventually absorbed into the stroma of the ovary. The role of the CL during pregnancy is discussed in chapter 5.

THREE

Fertilisation

Unless they encounter each other and complete the process of fertilisation, the gametes – both sperm and oocyte – are destined to die quite quickly (sperm within 45 hours, oocytes within 24 hours). The oocyte, with its complex cytoplasm, is the only cell in the body which has the potential to become a new human being, but without the contribution of the sperm genome its development cannot proceed. Fertilisation, the union of male and female gametes to form a **zygote**, is not a single event, but a continuum of subtle interactions and their outcomes.

3.1 Final preparation of the gametes

Before they are able to join to form a zygote, the sperm and oocyte must undergo final changes which enable fertilisation to occur. The process of gametogenesis (chapter 2) has resulted in the production of sperm from the seminiferous tubules of the testis, and oocytes in the dictyate stage of developmental arrest.

Epididymal maturation of sperm

The epididymis (figures 2.4 and 3.1) is a long, much coiled duct lined with columnar epithelium which has both secretory and absorptive functions. When the sperm enter the head of the epididymis from the testis, they are immotile and unable to fertilise an oocyte. During their passage along the length of the epididymal duct, the sperm acquire twitching movements at first, followed by full motility with active forward progression. The mature sperm are stored in the tail of the epididymis until ejaculation. A current is maintained through the duct by active transport of ions across the epithelial cells. The development of sperm motility is accompanied by changes in the composition of the sperm surface membrane and other ultrastructural modifications. These are mainly brought about by subtle changes in the fluid in the lumen of the duct due to absorption and secretion by the epithelial cells. Spermatogenesis is a continuous process which lasts throughout adult life, and sperm are reabsorbed by the cells of the epididymis if ejaculation is not regular. Epididymal maturation of sperm is dependent upon androgen secretion by the Leydig cells of the testis.

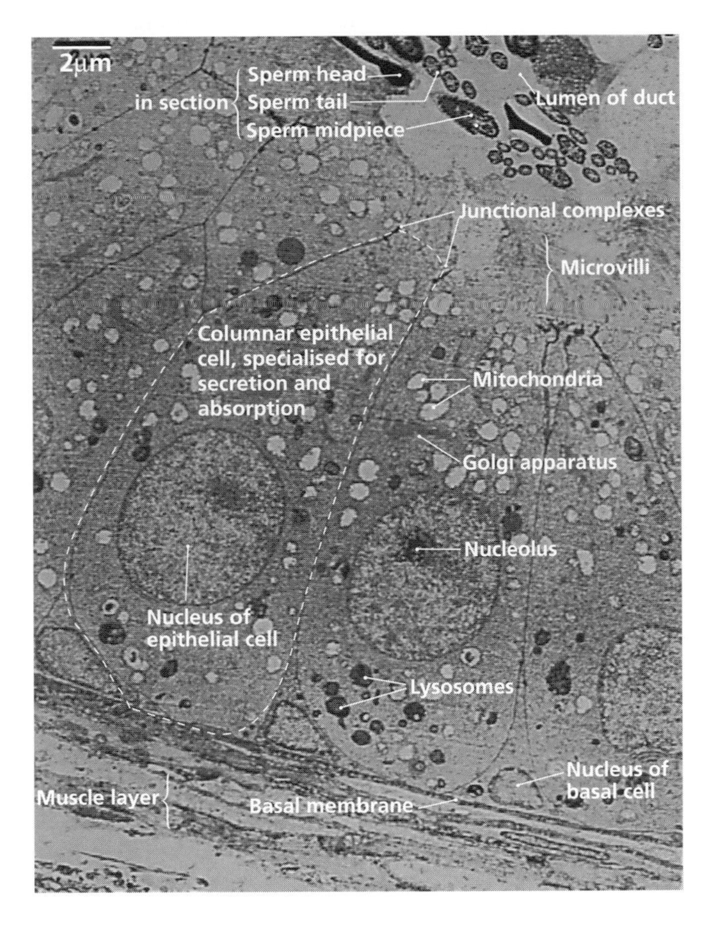

Figure 3.1 Electron micrograph of epididymal epithelium. Transverse section of the epididymal duct of a mammal. The epithelial cells lining the duct are specialised for both absorption and secretion. Maturing sperm can be seen at the top of the picture (x1500). (Electron micrograph, LM Baggott.)

Ejaculation and the journey of the sperm

During ejaculation, mature sperm from the tails of the epididymides pass into both vasa deferentia. At the same time, the prostate gland, Cowper's gland and seminal vesicles contract, releasing their secretions (together known as **seminal fluid**) into the vasa deferentia (figure 3.2). These also contract, and sperm are squeezed through the vasa deferentia where they mix with the seminal fluid. The **semen** (seminal fluid plus sperm) collects in the ejaculatory bulb of the urethra, and contraction of the part of the urethra in the penis causes emission of the semen.

In normal sexual intercourse, or **coitus,** about 5 ml of semen, containing around 300 000 spermatozoa, is ejaculated at the top of the vagina and around the cervix. Within a few minutes of intercourse, sperm can be found within the uterus. This is the beginning of a remarkable journey, about which relatively little is known. It has been estimated that less than 1 in 10 000 sperm (around 200) reach the site of fertilisation in the Fallopian tubes. Recent research suggests that sperm may be helped towards their goal by moving within cilia-lined 'tracks' in the lining of the uterus, but the high degree of specialisation for motility seen in sperm cells suggests that it is largely by their own propulsion that they move toward the Fallopian tubes (figure 3.3).

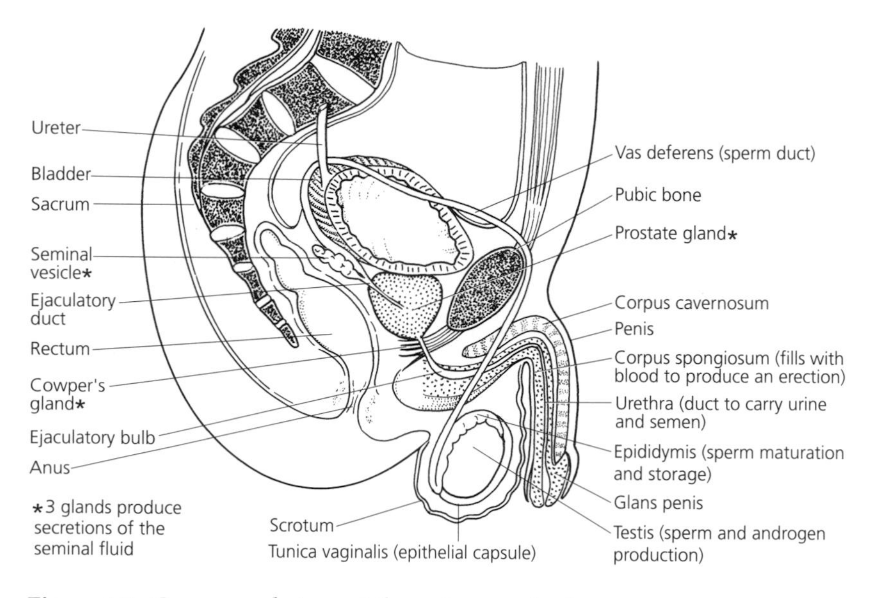

Figure 3.2 Gross morphology of the male reproductive system.

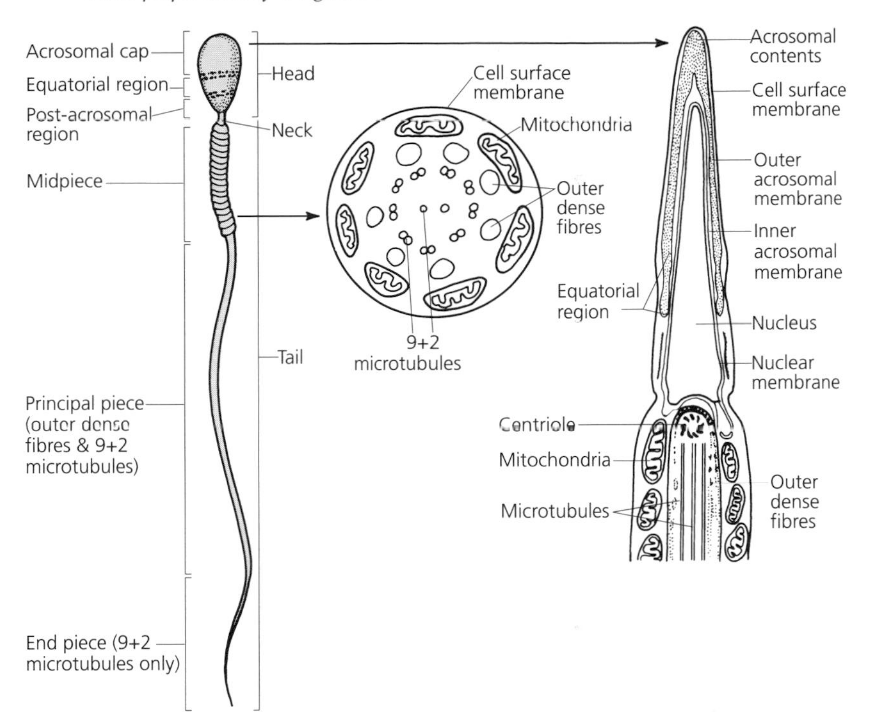

Figure 3.3 Ultrastructure of a mature spermatozoon, showing detail of the ultrastructure of the head and midpiece.

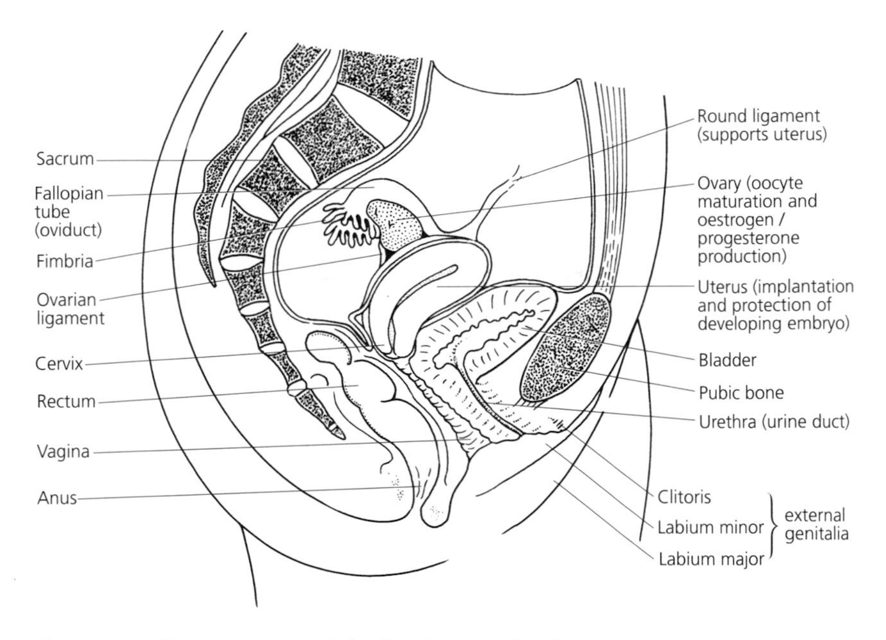

Figure 3.4 Gross anatomy of the female reproductive tract.

Sperm and oocyte in the Fallopian tube

In addition to the two ovaries, the female reproductive tract consists of a series of tubes and chambers adapted for internal fertilisation and the protection and nourishing of the developing fetus (figure 3.4).

The Fallopian tube is lined with a simple epithelium consisting of ciliated and secretory cells (figure 3.5). These two cell types are critical for the survival of both sperm and oocyte, for fertilisation and for early embryological development. Around the time of ovulation, the epithelial tissue of the Fallopian tube is very active. The cilia are at their largest and strongest at this time (see chapter 2), which is at mid-cycle. They function to mix the secretions in the lumen of the duct and to create a current by which the oocyte is transported. In the secretory cells, the synthesis of glycoproteins increases. These accumulate within vesicles which discharge their contents by exocytosis, thereby contributing to the fluid of the Fallopian tube. Other components of the oviductal fluid include pyruvic acid, glucose, albumin, immunoglobulin, ions and water. This fluid is derived from the blood plasma, the peritoneum, the uterus and the ruptured Graafian follicle.

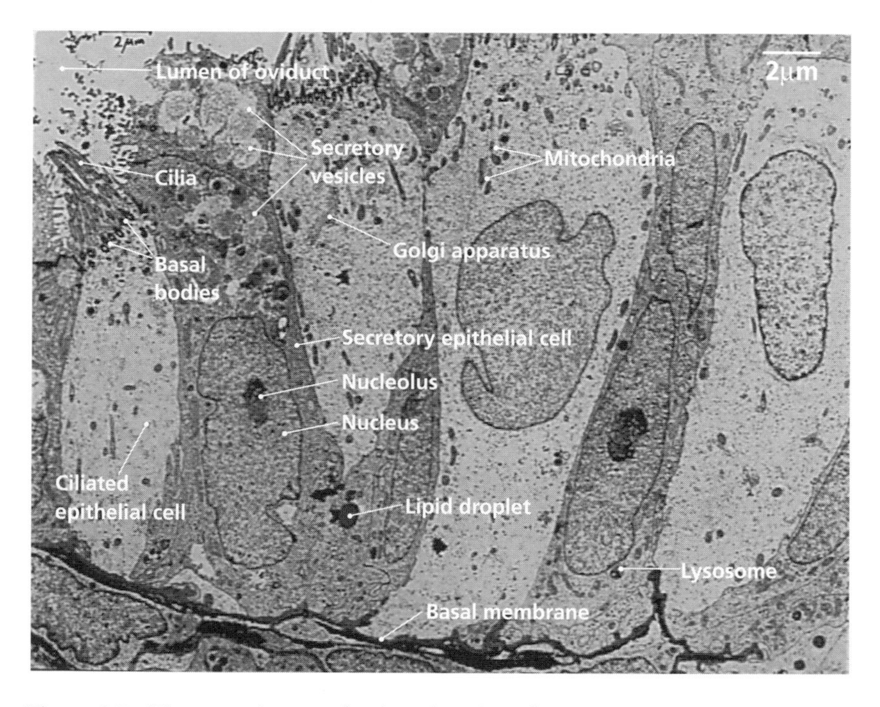

Figure 3.5 Electron micrograph of oviductal epithelium. Transverse section of the oviduct (Fallopian tube) of a mammal. Two types of epithelial cells can be seen lining the duct: ciliated (lighter) and secretory (darker) (x1500). (Electron micrograph, LM Baggott.)

Capacitation of sperm

In order to interact with the oocyte and to fertilise, the sperm must undergo a further process of modification by secretions of the Fallopian tubes. This is called **capacitation**. It takes up to six hours and involves two separate processes. The first is the acquisition of **hyperactivated motility**. During their journey toward the egg in the furthest section of the Fallopian tube, the sperm swim with a characteristic 'corkscrew' motion of the tail. After a suitable period in the proximity of the oocyte, the sperm become hyperactivated, taking on a different pattern of motility, often described as 'star-spin' (figure 3.6).

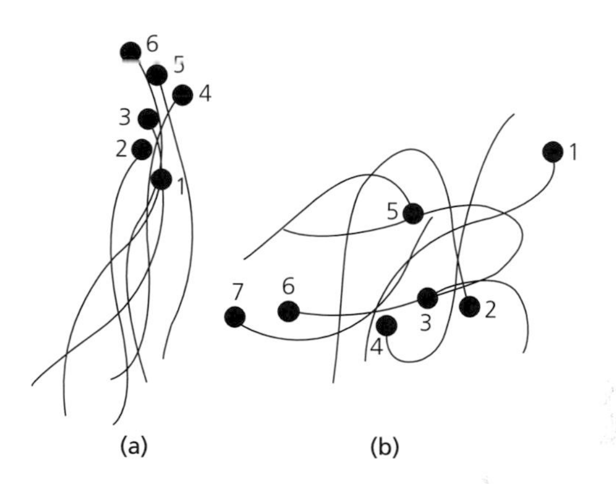

Figure 3.6 Flagellar movements of epididymal and hyperactivated sperm. The position of each sperm is shown after a time lapse of approximately 0.05 seconds. (*a*) Sperm from the epididymis. The figure-of-eight flagellar beating pattern is seen in uncapacitated sperm. (*b*) Hyperactivated sperm. This pattern, variously described as 'whiplash' or 'star-spin', is seen in capacitated sperm at or near the site of fertilisation. (After Yanagamachi, 1981.)

The reason for this change in swimming pattern is not fully understood. Some researchers believe that it helps sperm to swim better in the more viscous fluid surrounding the oocyte in the Fallopian tube, whilst others say it is a response to a chemical secreted by the egg when it is ready for fertilisation (**chemotaxis**).

The second capacitational change which sperm undergo in the Fallopian tubes is the facilitation of the **acrosome reaction**. This reaction occurs as the sperm encounters the oocyte, but first a period of capacitation must take place during which changes occur in the glycoprotein and lipid content of the sperm cell surface membrane. If this does not occur, the acrosome reaction does not take place, and the subsequent events of fertilisation cannot follow.

3.2 Cellular events of gamete interaction

The acrosome reaction

The first barrier the fertilising sperm encounters is the mass of cumulus cells (the **cumulus oophorus**) which surrounds the oocyte from ovulation. These cells are loosely packed, and are held together by a matrix containing **hyaluronic acid**. When the sperm comes into contact with this mass of cells, the acrosome reaction is initiated, heralded by a change in the permeability of the membrane covering the acrosome on the sperm head (figure 3.7). As a consequence the inner and outer acrosomal membranes 'point fuse', and holes appear in the covering of the acrosome through which its contents begin to diffuse. Among the enzymes contained in the acrosome is **hyaluronidase** which digests the hyaluronic acid, thus allowing the sperm to pass through to the **zona pellucida** beneath the cumulus layer.

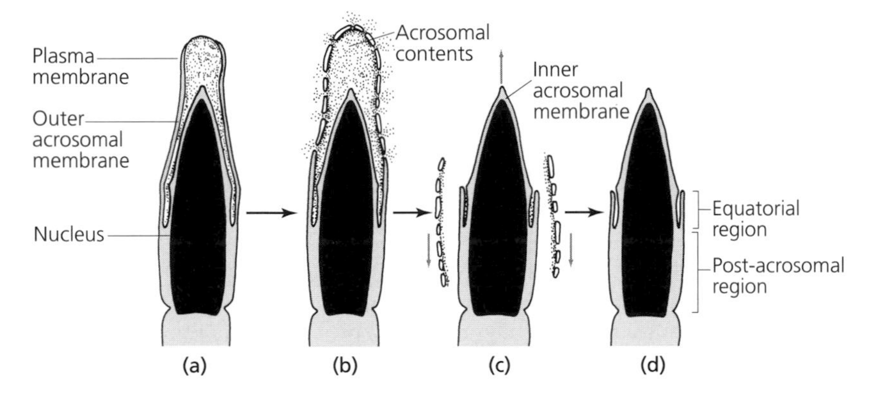

Figure 3.7 Progression of the acrosome reaction. (*a*) Before the acrosomes reaction. (*b*) Changes in the permeability of the membranes to Ca^{2+} causes the cell surface membrane to 'point fuse' with the outer acrosomal membrane. This allows the acrosomal contents to diffuse out, releasing enzymes to help sperm penetrate the outer coats of the oocyte. (*c*) The fused membranes are sloughed off, leaving the inner acrosomal membrane exposed. (*d*) Physiological changes occur in the equatorial region of the sperm head, rendering it capable of fusion with the oocyte surface membrane.

Some researchers believe that dispersal of the cumulus oophorus is not necessary for the sperm to bind to the zona pellucida, but other evidence shows that antibodies to hyaluronidase effectively block cumulus dispersal, and that subsequent fertilisation is prevented.

Penetration of the zona pellucida

The zona pellucida is composed of three glycoproteins, ZP1, ZP2 and ZP3. The last has been identified as a **sperm receptor**, and the exposed inner acrosomal membrane on the sperm head contains a receptor protein with which ZP3 can interact. The sperm makes a neat penetration slit in the zona pellucida by further enzymatic activity, possibly by an acrosomal protease called **acrosin**, and passes through the zona pellucida by propulsive force from the sperm tail.

Interaction of sperm with oocyte surface membrane

Having penetrated the zona, the sperm emerges into the **perivitelline space**. This presents no real impediment to its progress, and it swims forward vigorously before flattening its head against the oocyte membrane. This membrane contains numerous **microvilli**, and it is with these that the equatorial region of the sperm cell surface membrane initially binds. At this point, the sperm tail is still beating strongly, and this causes the whole oocyte to rotate within the zona pellucida. The sperm membrane gradually fuses along its length with the underlying oocyte membrane. Throughout this time the **cortical granules** of the oocyte exocytose their contents into the perivitelline space as a result of a sharp increase in the concentration of calcium ions (Ca^{2+}) in the oocyte cytoplasm. The substances released in this cortical granule reaction interact with the receptors on the zona pellucida to prevent the binding of further sperm. Thus it is thought that the cortical granule reaction plays a major part in the **block to polyspermy**, or penetration of the oocyte by more than one sperm. As soon as it is incorporated into the oocyte cytoplasm, the sperm

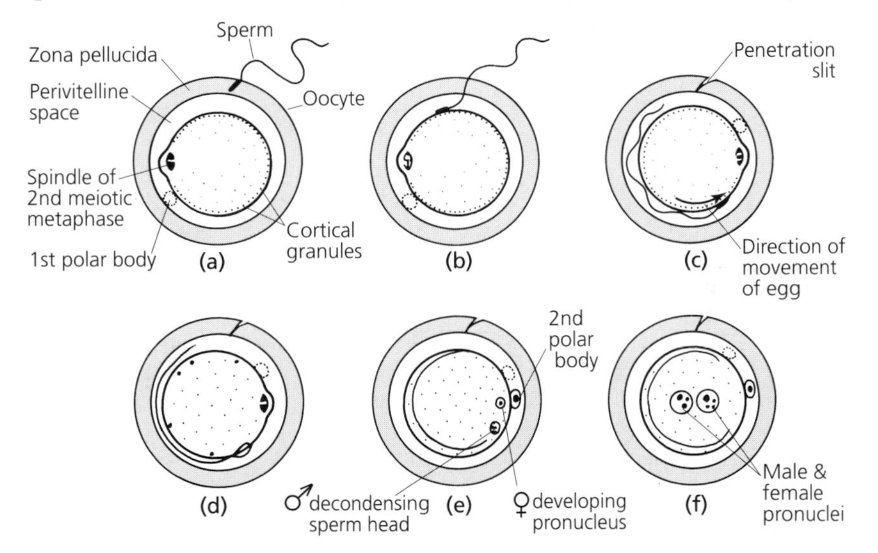

Figure 3.8 The process of fertilisation. (*a*) The fertilising sperm contacts the zona pellucida. (*b*) The sperm tail beats vigorously, causing the oocyte to rotate within the zona pellucida, until the tail is brought within the perivitelline space. (*d*) The cortical granules undergo exocytosis into the perivitelline space (the cortical granule reaction), and the sperm head and oocyte surface membranes fuse. (*e*) The second meiotic division of the oocyte is completed, and the second polar body is formed. (*f*) The male and female pronuclei form.

nuclear membrane breaks down, and the tightly packed chromatin in the sperm head begins to decondense.

Oocyte activation

The increase in oocyte intracellular calcium ions not only brings about the cortical granule reaction, but also stimulates the cell's metabolism, with consequent rises in the rate of cellular respiration and protein synthesis. When these changes have occurred, the oocyte is said to be **activated**. These events, together with the incorporation of the sperm head, cause the resumption of meiosis – which was arrested at metaphase II. The zygote proceeds quickly through the rest of the division, with the resulting extrusion of a second **polar body**. The first polar body simultaneously completes its second meiotic division (figure 3.8).

3.3 Chromosomal events leading to the first cleavage

Within four hours of fusion, each set of haploid chromosomes becomes surrounded by membranes and are visible as the male and female **pronuclei** (figure 3.9).

The male pronucleus is slightly the larger, but both pronuclei contain several prominent nucleoli. Gradually the two pronuclei move to a central position within the zygotic cytoplasm, and enter the S phase of the first embryonic cell cycle. After 18–21 hours, the pronuclear membranes

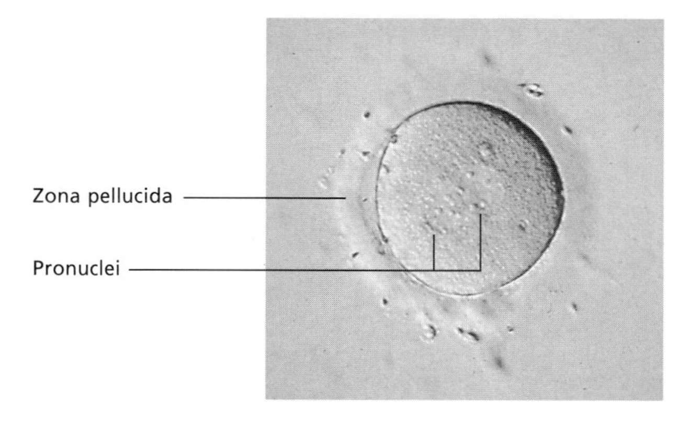

Figure 3.9 Photomicrograph of pronuclear stage.

break down, and the chromosomes condense and intermingle during mitotic prophase. The spindle apparatus of metaphase appears and the zygotic chromosomes take up their position on the equator of the cell for the first time. The end of the subsequent phases of this mitotic division signals the arrival of **syngamy**, the final stage of fertilisation, in which the haploid gametic chromosomes have come together in a united diploid genome. The oocyte, with the sperm, has become an embryo (figure 3.10).

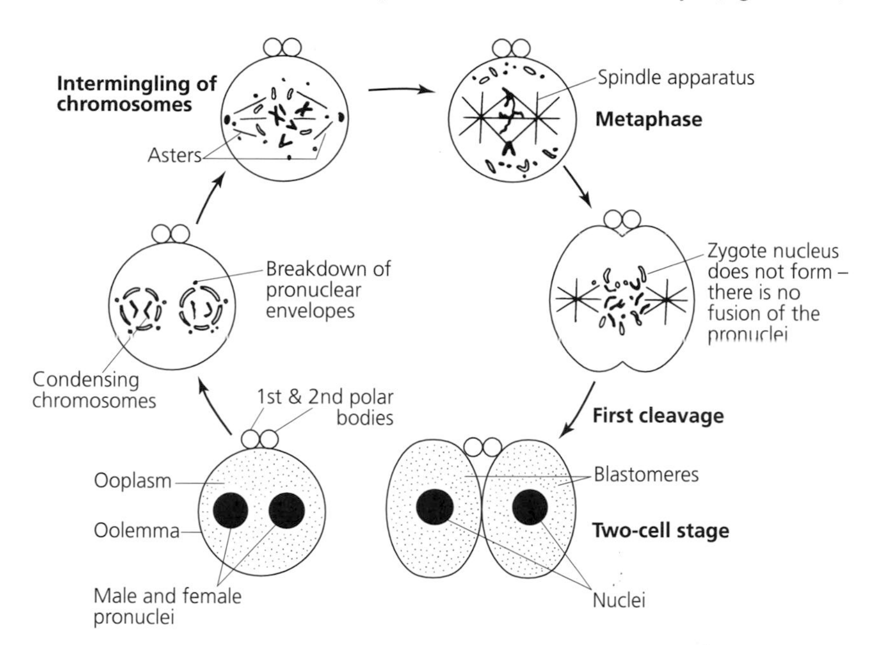

Figure 3.10 The first cleavage division, showing association of the male and female pronuclei. Note that the zygote nucleus does not form after the first cleavage division, and the male and female pronuclei do not fuse. The maternal and paternal genomes are associated within a single nucleus for the first time in the blastomeres of the 2-cell stage embryo. For a photomicrograph of a 2-cell embryo, see figure 4.1.

3.4 Sex determination

In humans, gender is determined chromosomally. Sperm may carry an X or a Y chromosome in addition to the 22 autosomes, whereas oocytes normally carry 22 autosomes plus an X chromosome. Therefore inheritance of gender depends on whether an X-bearing or a Y-bearing sperm fertilises the oocyte. This has been known since around 1959, but the underlying mechanism has only recently been elucidated. It is now known that male development requires the presence of a specific gene called SRY, which is located on the short arm of the Y chromosome. This chromosome contains very few genes, but this **testis determining gene** is one exception. It codes for a protein (**testis determining factor, TDF**) which stimulates the cells of the gonadal ridge to differentiate into seminiferous tubules and interstitial (Leydig) cells. In about the ninth week of fetal life, the Leydig cells develop to release testosterone into the circulation, where it combines with the androgen receptor molecules in the cells of the developing reproductive tract (see chapter 4). The androgen receptor/testosterone complex passes to the nuclei of these embryonic cells where it activates genes associated with the development of the various tissues of the male tract: epididymis, vas deferens, prostate gland and so on. In an XX embryo, the absence of the SRY gene, and hence the absence of the cascade of male sexual development, allows the reproductive system to develop in its inherent direction towards the female.

Experiments with mouse embryos have reinforced this explanation of the genetic basis of sex determination. The mouse gene Sry is equivalent to the human SRY. Hundreds of fertilised mouse oocytes were injected with Sry gene sequences, and after incubation 158 embryos were obtained. The majority of these did not incorporate the Sry gene into their genetic material and developed into either normal XY males or normal XX females. In two cases, however, testes were seen in embryos with XX chromosomes. Tests showed that these embryos had the Sry gene, and that development of an apparently normal XY male reproductive tract had taken place. One of the two embryos developed to birth, and was seen to have normal male external genitalia. This animal showed normal sexual behaviour when caged with females, but no litters were born to the females mated with the XX male. Subsequent histological examination of the testes of this male showed that although tubules with Sertoli cells and Leydig cells were present, there was no sign of spermatogenesis.

Unlike the Y chromosome, the human X chromosome contains the genes for at least 90 known characters. These genes have been highly conserved during evolution, so if a character is sex-linked in one species, it tends to be sex-linked in all species. The female survives the effects of a double dose of genes on the paired X chromosome because of the phenomenon of **X-inactivation**. This is the process by which almost all the genes on one of the two X chromosomes in somatic cells are suppressed. Which chromosome in a cell is inactivated is determined at random early in embryonic life, and thereafter all the cell's progeny will follow suit. Evidence for this comes from tortoiseshell cats which have a mixture of black hair, produced by the dominant gene B, and yellow hair, produced by the recessive gene b. This gene is X-linked. A female cat heterozygous for this character has mottled black and yellow hair, showing random X-inactivation in her skin cells.

Male tortoiseshell cats are rare – most are XXY and sterile. Primordial germ cells are exceptions to the rule of random X-inactivation, as are any cells with a defective X chromosome which is automatically inactivated, and the cells of individuals with more than one X chromosome. This topic is covered more fully in *Heredity and human diversity* by Stephen Tomkins (CUP, 1989).

F O U R

Implantation and pregnancy

4.1 Significance of viviparity

The oocyte is one of the biggest cells in the human body, and this is due mainly to the relatively large amount of cytoplasm which it contains. A major constituent of this **ooplasm** is yolk which is the main nutritional source for the embryo in its initial stage of development. Yolk is a mixture of lipids and proteins which provide the energy required for what is a very rapid period of biological change. The oocytes of all animals contain yolk, but the relative amount depends upon the degree of isolation of the embryo from other sources of nutrition. Birds and reptiles, for example, must complete all of their embryological development from zygote to hatchling within the confines of the egg shell, so they invariably have very yolky or **teleolecithal** eggs. Those species whose embryos derive their nourishment entirely from the egg yolk are described as **oviparous**.

The eggs of mammals contain relatively little yolk compared with the eggs of other vertebrate species. However, yolk is present in sufficient quantity to sustain the development of the early embryo through the period of **cleavage**. This term is used to describe the first few cell cycles of a multicellular organism developing from a single fertilised oocyte. Mammalian eggs are termed **alecithal** (figures 2.9 and 2.10).

In placental mammals, including of course humans, cleavage takes place as the embryo passes down the Fallopian tube towards the uterus. During this period, the embryo draws upon the reserves of yolk in the dividing cells. After implantation has taken place in the uterus and until birth, energy and raw materials come to the developing embryo from the maternal circulation. In this sense, the embryo behaves as a parasite within the mother. Species which have evolved this strategy for nourishing their developing offspring are said to be **viviparous**. In evolutionary terms, viviparity with internal fertilisation means females of the species need produce relatively few female gametes, though the protection of the developing young within the mother's body requires special structures, such as uterus and placenta. The mammary glands, or breasts, have also evolved to support this reproductive strategy. Although not part of the reproductive system as such, they are clearly designed to play a very important part in post-natal parental care.

4.2 Development of embryo from fertilisation to implantation

Cleavage

Cleavage, as defined in section 4.1, represents the first cell cycles of embryonic life. However, although similar structurally and functionally to somatic cell division and binary fission in protozoans, the outcome of cleavage for the zygote is fundamentally different. The zygote is changed by a rapid series of mitoses and cytokineses from a unicellular fertilised oocyte into a multicellular complex of cells called **blastomeres**. These blastomeres differ from the zygote in both their contents and, importantly, their potential for development. Within four days of fertilisation, the cleaved embryo may contain only around 100 cells, but these cells are very active metabolically; approximate to the size of the somatic cells of the adult body; and the genes have begun to function independently, coding for proteins which are immunologically distinct from those of the mother, father or zygote. There is no increase in mass during the first few days following fertilisation: the sum of the masses of the blastomeres is usually just less than that of the zygote. This is due to some of the stored energy in the yolky deposits of the cytoplasm being used, as well as to the dispersal of waste products of metabolism. However, the ratio of nucleus to cytoplasm, which is very low in the zygote, is similar to that seen in ordinary somatic cells by the end of cleavage.

Totipotency

The constituent parts of the oocyte cytoplasm tend not to be displaced by the process of cleavage, but generally remain in the same position as at the start. In some animals, e.g. annelids, molluscs and many primitive chordates, most cytoplasmic organisation occurs after fertilisation and prior to the first cleavage division, and so the fate of the blastomeres is determined before cleavage. This pattern of development is called **determinative** or **mosaic development**. In other animals, notably the echinoderms and vertebrates, including humans, interactions between cytoplasmic components of the egg are delayed until cleavage has progressed far enough for cell movements and interactions to become possible. This mode of development is called **indeterminate** and single blastomeres taken from early cleavage stages can be made to give rise to whole embryos. The blastomeres are said to exhibit **totipotency**.

Early cleavage divisions

Just over 24 hours after fertilisation, the human zygote divides, or cleaves, for the first time, giving rise to a 2-cell embryo (figure 4.1a). This is followed at intervals of about 10–12 hours by successive cleavage divisions. At first these are synchronous, but after 4–5 cleavage divisions this synchrony disappears. Each mitosis is followed immediately by the S phase of the next cell cycle. This means that there is no G1 phase. The embryo, by now at about the 8-cell stage, enters the uterus at about 2.5 days, but by then a remarkable and significant change has occurred. The spherical blastomeres have flattened against each other producing a larger area of cell-to-cell contact, and intercellular communication begins. This process is called **compaction**, and the embryo is known as a **morula** (figure 4.1c). The significance of this stage is that for the first time some cells of the embryo are different from others. This confers **polarity** on the embryo, and marks the beginning of the process of **differentiation**.

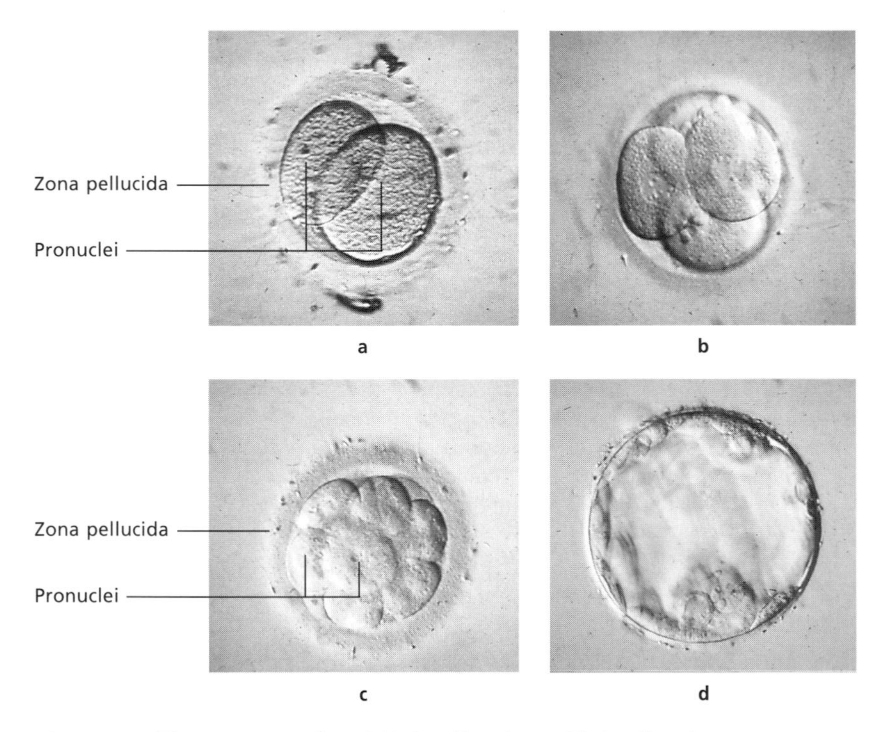

Figure 4.1 Photomicrographs of (*a*) 2–cell embryo, (*b*) 4–cell embryo, (*c*) morula, and (*d*) blastocyst. In the blastocyst, the single layer of cells destined to become the extraembryonic membranes (trophectoderm) can be seen around the periphery flattened against the zona pellucida. The inner cell mass of smaller cells eventually develops into the embryo itself.

Blastocyst formation

A fluid-filled cavity, the **blastocoel**, appears within the cells of the morula, and as the cells continue to divide the cavity enlarges, eventually producing a hollow sphere consisting of two distinct types of cell. This structure is called a **blastocyst** (figure 4.1d). Around most of its periphery is a single layer of flattened epithelial cells which is destined to become the extraembryonic membranes, and a group of smaller cells which eventually gives rise to the embryo itself (figure 4.2).

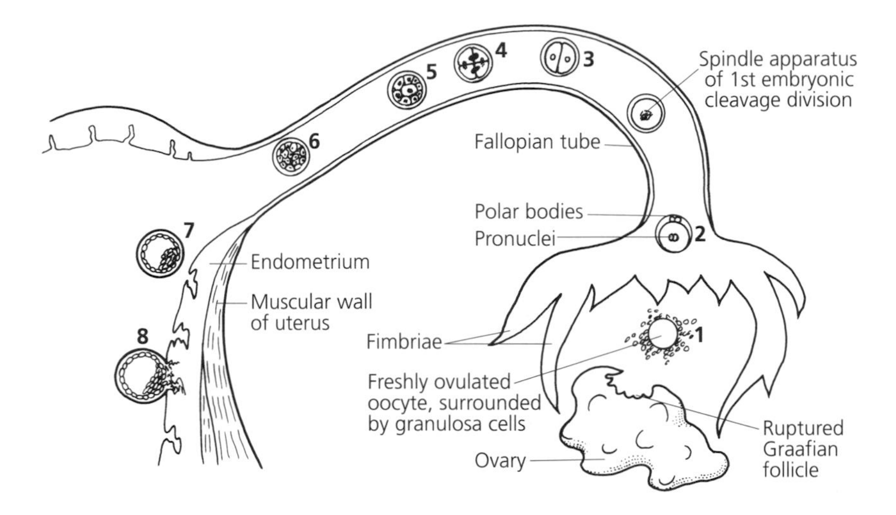

Figure 4.2 Diagram of cleavage in the female tract. *1* The oocyte, surrounded by the cumulus mass of cells, is ovulated from the ovary. *2* Fertilisation takes place in the ampulla of the Fallopian tube, and the pronuclear stage follows. *3* The first cleavage division follows about 1.5 days after ovulation. *4–6* In the next 24 hours, the embryo completes its journey along the Fallopian tube, during which time further cleavage divisions and compaction occur, followed by formation of the morula. *7* The embryo enters the uterus about 3 days after ovulation, and blastocyst formation occurs at 4 days. *8* The blastocyst commences implantation 8–13 days post-ovulation.
(After Baggott LM (1992) *J. Biol. Ed.* **26(3)**:171–177.)

4.3 Implantation

Blastocyst hatching

During cleavage, the zona pellucida serves to keep the cleaving embryo from fragmenting and to stop it implanting before it reaches the uterus. Upon reaching the normal site of implantation in the uterus, however, it now becomes an impediment to this process. By about 5 days after fertilisation, the blastocyst ‘hatches’ from the zona pellucida by squeezing through a hole created by enzymatic action (figure 4.3).

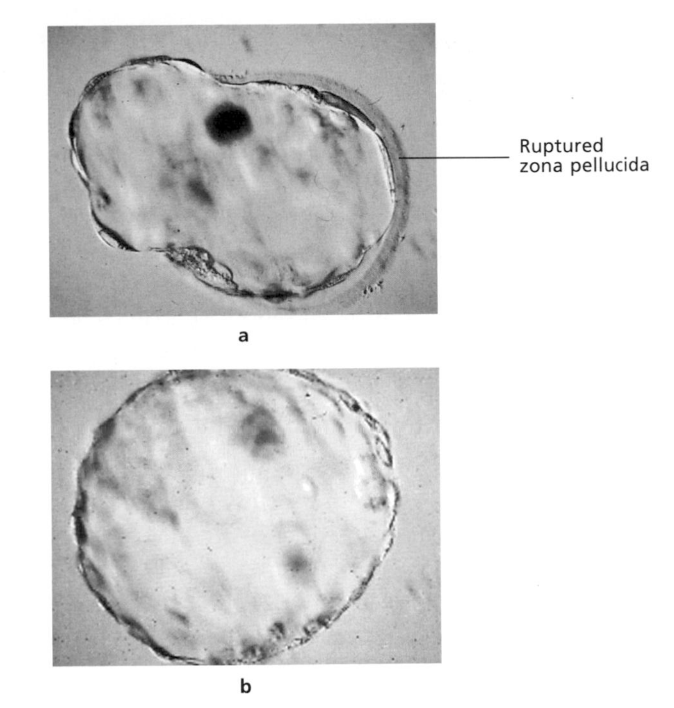

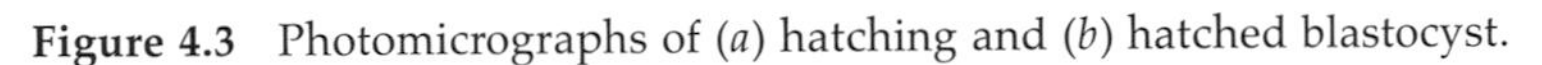

Figure 4.3 Photomicrographs of (*a*) hatching and (*b*) hatched blastocyst.

Implantation of blastocyst

Implantation is that point of embryonic development which brings the embryo into intimate association with the circulation of the mother, to ensure its survival. Now free of all its original investing layers, the blastocyst engages with the **endometrium**, to which it sticks. The cells destined to form the embryo itself are those nearest to the endometrial cells. Under the influence of progesterone secreted by the corpus luteum, the cells of the endometrium differentiate into actively secreting cells. Glands and blood vessels around the implantation site in the endometrium enlarge, and it is thought that their secretions may contribute growth factors to support the implanting embryo.

Implantation in an abnormal site results in what is termed an **ectopic pregnancy**. This can be extremely dangerous for the woman, since an implanting embryo and its associated tissues are highly invasive and this can cause blood vessels to rupture with risk of haemorrhage. The symptoms may include abdominal pain and/or vaginal bleeding. When this occurs, surgery may be necessary to remove the developing embryo and any surrounding tissues damaged by the attempted implantation.

Invasion of the endometrium by the outer layer of embryonic cells

Some of the cells of the outer layer of the embryo divide rapidly to form a **syncytium**, which is a region of a cytoplasm with many nuclei but no dividing cell surface membranes. This syncytial tissue is highly invasive, pushing its way into the uterine wall. In this way, the embryo eventually becomes completely buried in the wall of the uterus. A period of rapid growth and development now begins for the embryo, characterised by differentiation of the embryonic cells and the appearance of the **amniotic cavity** at about 7 days post-fertilisation. This cavity subsequently fills with a secretion, the **amniotic fluid** (see section 4.4).

Role of hormones in implantation

The endometrium is prevented from breaking down by the action of progesterone produced by the corpus luteum. If implantation does not occur, and a high proportion of cleaving embryos do fail to implant, the CL degenerates by about 13 days after ovulation. Successful implantation, however, results in production by the outer embryonic cells of the hormone **human chorionic gonadotrophin (hCG)**. This acts upon the CL, supporting its function and thereby maintaining the supply of progesterone. HCG is detectable from 8 days post-ovulation, and this forms the basis of commercially available pregnancy testing kits. Function of the CL is vital to the maintenance of the pregnancy in the first 11–12 weeks, but after that the placenta takes over the production of steroid hormones and the CL regresses to become a **corpus albicans**. HCG also has an important effect on the testes of the male fetus and fetal adrenal gland. It stimulates production of testosterone which in turn allows differentiation of the male urogenital tract, hypothalamus and cerebral hemispheres.

4.4 Functions of the placenta and extraembryonic membranes

Development of the placenta

As a consequence of implantation, the blood vessels of the endometrium are eroded to form blood-filled cavities in which structures called the **chorionic villi** develop. These are compartments with a specialised arterial blood supply between the fetal and maternal tissues. These arteries soon become invaded by cells of the developing placenta and become dilated. The blood flow in the chorionic villi is high but of low resistance, so they are covered with a fast turnover of blood with a very similar oxygen and carbon dioxide content to that of the maternal arteries.

Functions of the placenta

The placenta is responsible for the nutrition, gas exchange, excretion, anchorage of the fetus, and its protection from immunological attack. It also serves as an endocrine gland and as a heat exchanger. It increases in size and weight with the advancing pregnancy, but not at the same rate as the fetus.

The placenta is a very important endocrine organ, and its output of steroid (oestrogens and progesterone) and non-steroid (hCG, human placental lactogen (hPL) and prolactin) hormones increases as the pregnancy progresses. Progesterone is the most important hormone for the maintenance of the pregnancy, by preventing the breakdown of the decidual lining of the uterus and reducing the muscular action of the uterine wall (**myometrium**). Oestrogen is produced by the placenta in increasing quantities, and stimulates the growth of both the uterus and breasts. HPL and prolactin are important for the development of the breasts, and, while hPL may have other functions, prolactin is important in the stimulation of lactation. Before birth, this is inhibited by oestrogen produced by the placenta.

Role of the amniotic fluid

The amniotic fluid is produced from the amnion and functions as a shock absorber for the embryo, and also prevents it from drying out. After the kidneys begin to function during the tenth week of fetal life, urine output makes a significant contribution to the amniotic fluid which increases to about 0.5 litres per day at the end of pregnancy. In the fourth to the sixth months of pregnancy (second **trimester**) the fluid has a composition resembling that of fetal tissue fluid, but in the third trimester (seventh to ninth months) it approximates more closely to fetal urine. Just before birth the fetus swallows up to 0.5 litres of amniotic fluid per day, but the net increase in amniotic fluid at this time is around 5 ml per day.

4.5 Formation of twins

The formation of dizygotic twins, which result from the fertilisation of more than one oocyte, was outlined in section 2.7. Because they are the result of two oocytes and two sperm, they have separate genetic identities and are no more alike than they would be to any other of their brothers or sisters. They may, of course, be of different sexes. These non-identical twins have completely separate amnions, chorions and placentae, but fusion of the placentae is common, particularly when the implantation sites are close.

Monozygotic twins develop from the same fertilised oocyte, and are genetically identical. They are less common than dizygotic twins, occurring in 30–40 pregnancies per 10 000. There are three types depending on when the zygote divides in two: at morula stage, when they have separate placentae, chorionic and amnionic cavities; at the blastocyst (inner cell mass) stage, when they share placenta and chorion but have separate amniotic cavities (this is the most common time for separation); and after the formation of the amniotic cavity, when they share amnion, chorion and placenta. This latter situation, which occurs in less than 3% of monozygotic twinning, arises as a result of gastrulation (see section 4.6) beginning at two separate sites on the embryonic disc. These twin embryos float in the same amniotic cavity, and the umbilical cords are quite likely to become entangled. It is believed that conjoined or Siamese twins arise when gastrulation begins simultaneously in this way, but the two body axes become so close that they fuse.

Multiple pregnancies are always regarded by obstetricians as potentially complicated, and are closely monitored. The woman is rarely allowed to go beyond 38 weeks of pregnancy because of the risk of placental failure, and she is always advised to have her delivery in hospital. Higher order births are even rarer, in the absence of fertility treatment, than twins. Triplets occur naturally in about 1.3 per 10 000 pregnancies, and quadruplets in about 1.7 per million.

4.6 Outline of fetal development to birth

It is beyond the scope of this book to go into detail on the highly complex subject of human embryology: accounts of this can be read elsewhere. The gestation period is on average 266 days, or about nine months. A pregnancy is always reckoned from the first day of the woman's last menstrual period, and can be divided into three approximately equal trimesters. During the first trimester the zygote grows and develops into a fetus, reaching a length of about 32 mm. Its mass increases by over 500 times, and its cells and organs differentiate and become organised into a recognisably human form. The second and third trimesters are characterised by growth of the fetus, placenta and uterus, and preparation for birth.

Gastrulation

Gastrulation is a very important process by which all the tissues which make up the fully developed individual are founded, and the basic body plan is laid down. It is a series of integrated and coordinated cell movements which bring about the rearrangement of the cells of the blastocyst. During gastrulation the three germ layers of the body are established in their respective positions: **endoderm** and **mesoderm** within the embryo and **ectoderm** outside it. It involves movements of individual cells, groups and sheets of cells, and folding of the layers produced by these movements (figure 4.4).

By about 15 days since fertilisation, a structure called the **primitive streak** appears in the embryo. During gastrulation, the cells that go on to make up all the tissues of the embryo migrate through the primitive streak to form the **mesoderm** and **endoderm**. Thus the process of gastrulation establishes not only the layers of the body, but also the main body axis and the bilateral symmetry of the embryo.

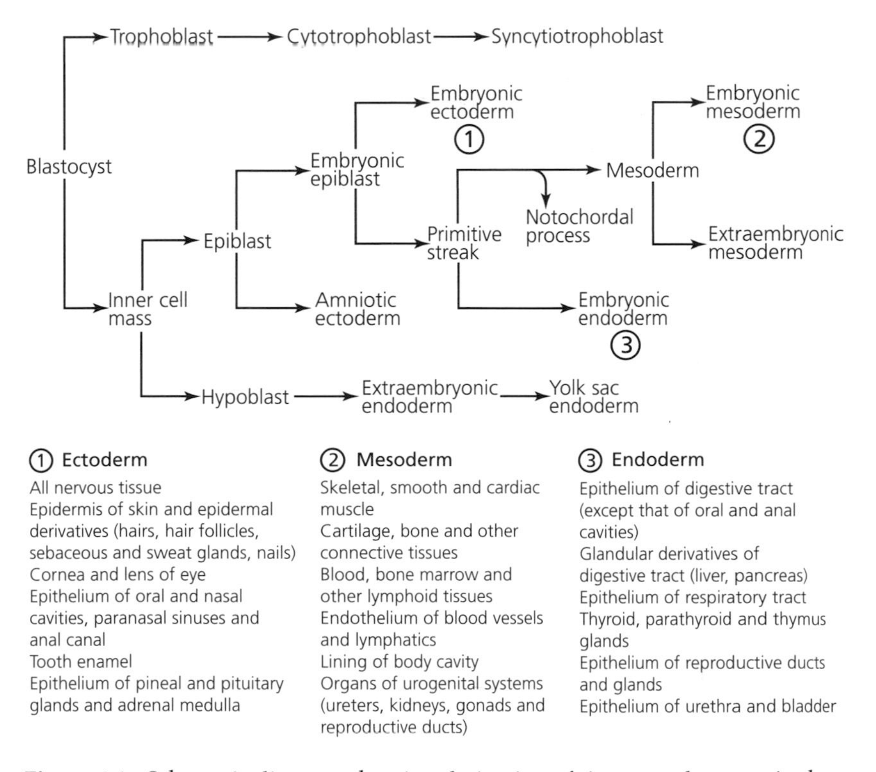

Figure 4.4 Schematic diagram showing derivation of tissues and organs in the human embryo.

Source: Adapted from Elaine N. Marieb, *Human Anatomy and Physiology*, 2nd ed. Redwood City, CA: Benjamin/Cummings, 1992.

Organogenesis

The period which follows gastrulation is known as **organogenesis**, and during this period – 4–8 weeks post-fertilisation – specific tissues, organs and systems are formed (figure 4.5). Prior to this, immediately following gastrulation, two important structures are laid down. The first of these are the **somites**. These are blocks of tissue which develop either side of

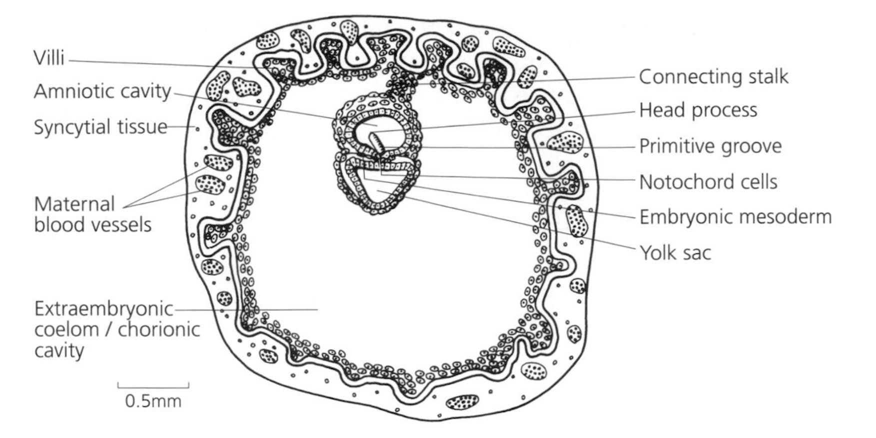

Figure 4.5 Arrangement of embryonic and extraembryonic structures following gastrulation.

the **notochord**, a rod of cells which directs the formation of the neural tube. The somites eventually give rise to the vertebral column, skeletal muscles and dermis. The other precursor is the **neural plate**, which first appears as a thickening of the ectoderm and eventually folds into the neural tube, the precursor of the central nervous system. A special population of cells, the **neural crest**, becomes detached from the neural tube, and the cells migrate to various other locations of the body where they develop into a variety of tissues.

The period of organogenesis, beginning with the formation of the neural tube and somites, is essentially complete by the end of the first trimester of pregnancy. Figure 4.4 shows the derivation of all the major organs of the body from the three primary germ layers.

4.7 Development of the reproductive system

By its very nature, the human life cycle has no clearly demarcated beginning and end, and any discussion of it could start at a number of different points. Aspects of human development concerned with reproduction are described in subsequent chapters, but a brief outline here of the formation and differentiation of the embryonic genital system will provide an introduction to the structure and function of the system in the adult. Development of the reproductive system is closely linked to that of the urinary system, and as early as 26 days after fertilisation a structure called the **mesonephros** differentiates from the primitive embryonic tissues. It is thought that this may function as a kidney during embryonic life, and it also plays an important role in the development of the male reproductive tract.

Formation of the genital ridges

The cells which give rise to the gametes are known as primordial germ cells (chapter 2), and they differentiate during the fourth week after fertilisation. During the fourth to sixth week of development, these as yet unspecialised cells migrate, multiplying continuously by mitosis, to a position within the embryo where their presence ultimately brings about the formation of the gonads (ovaries and testes). The arrival of the primordial germ cells at their destination stimulates the surrounding tissues to multiply and form strands of cells. These developing strands cause swellings – the **genital ridges** – which lie on either side of the midline of the embryo, between the embryonic kidney (mesonephros) and the vertebral column (figure 4.6). During the sixth week, new structures called the **Müllerian ducts** begin to form either side of the mesonephric ducts in both sexes. At this stage these presumptive gonads are undifferentiated and there is no visible difference between the sexes, although there may be subtle distinctions at the cellular level.

The tissues which provide the strands of cells consist of two distinct regions. After the sixth week of embryonic life these two regions develop differently in the female and male. This process is under genetic control and is discussed more fully in chapter 3. The cells eventually give rise to the tissues that nourish and control the development of the gametes. In the female these tissues are the **ovarian follicles**, and in the male the seminiferous tubules of the testis.

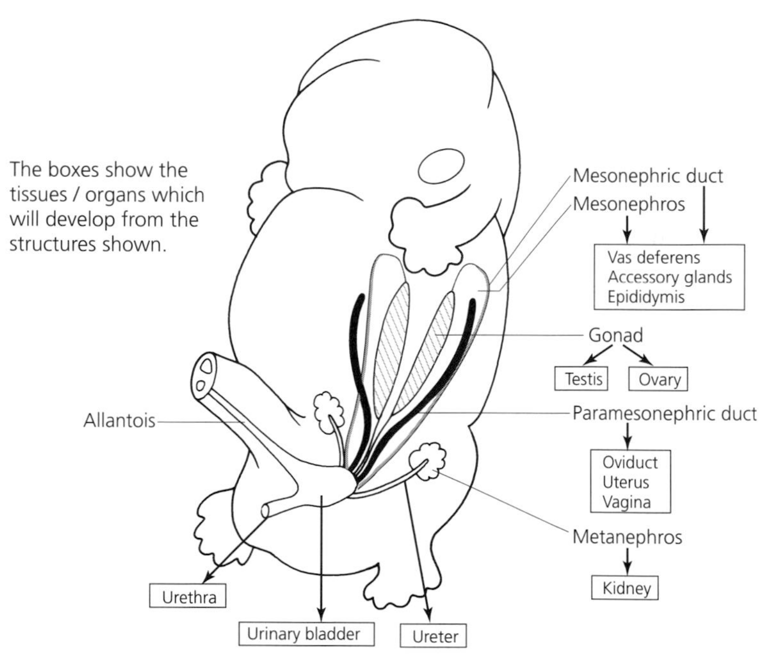

Figure 4.6 Formation of the genital ridges during weeks 5–6 of embryonic life.

Differentiation of the testis and ovary

During the seventh week the male and female reproductive tracts develop from these early embryonic tissues (figure 4.7). In the female they give rise to the Fallopian tubes, uterus and vagina. The developing testis in the male produces hormones which bring about the differentiation of the vas deferens, epididymis and accessory glands of the male urethra.

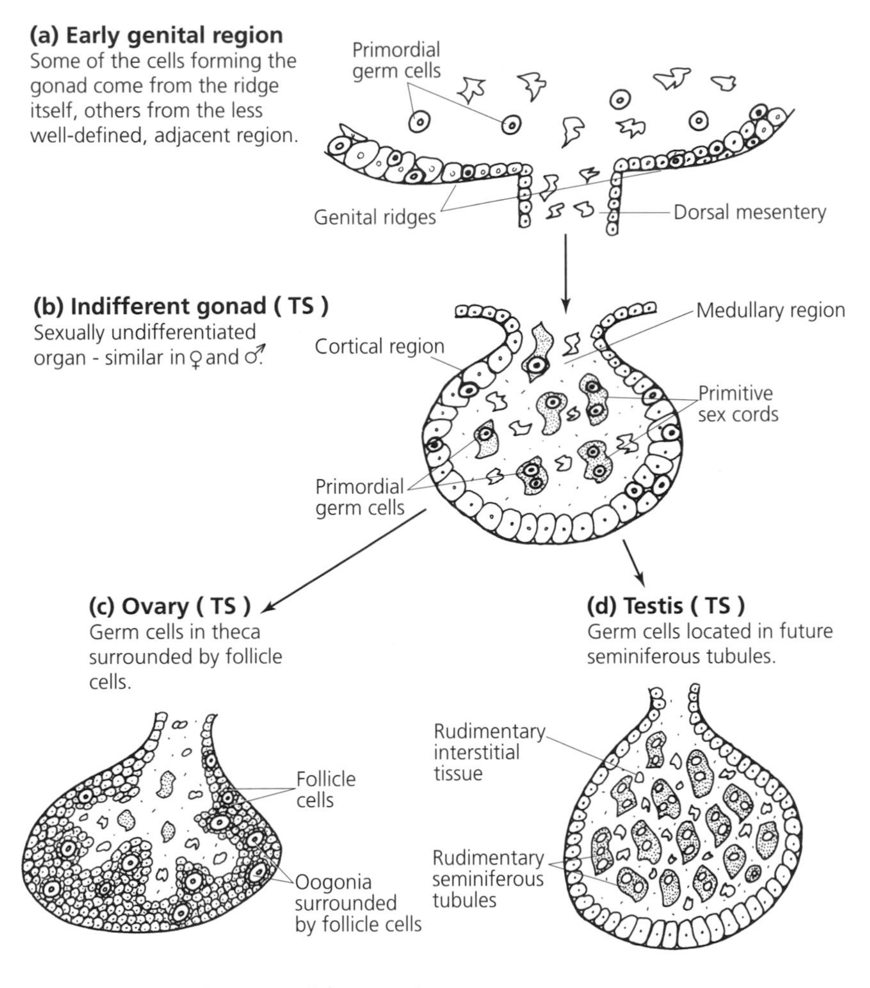

Figure 4.7 Development of the gonads.

The external genitalia start to differentiate at around the twelfth week of embryonic life. By sixteen weeks, the rudimentary phallus has developed into the penis in the male and into the clitoris in the female. The genital folds have formed the scrotum in the male and the labia in the female, and a space called the **urogenital sinus** in the embryo remains to form the vaginal opening in the female, but closes by week 14 in the male.

4.8 Preparation for extrauterine life

Although specialist perinatal care can enable premature babies and those of extremely low birth weight to survive, a child born below 500 g or after less than 24 weeks' gestation rarely lives. The main cause of death is failure of the baby to breathe. This is because the surface tension of the fluid within the alveoli of the lungs exceeds the muscular strength of the baby to inspire, with the result that the alveoli fail to expand. A substance called **surfactant** is produced at some time after 24 weeks which has detergent qualities that lower the surface tension of the fluid in the alveoli thus enabling the baby to breathe.

The lungs of the fetus *in utero* do not function in gas exchange, and the pulmonary arteries carry very little blood during fetal life. The umbilical vein carries oxygenated blood to the inferior vena cava which empties into the right atrium of the heart. A large proportion of the blood then passes through the **foramen ovale**, a hole which connects the atria in the fetal heart, and then passes to the left ventricle which pumps it via the aorta around the body. The rest of the blood which has not passed through the foramen ovale is pumped to the right ventricle and thence into the **ductus arteriosus**, another purely fetal structure which joins the aorta (figure 4.8). The changes which occur in the fetal cardiopulmonary system at the time of birth are discussed in chapter 5.

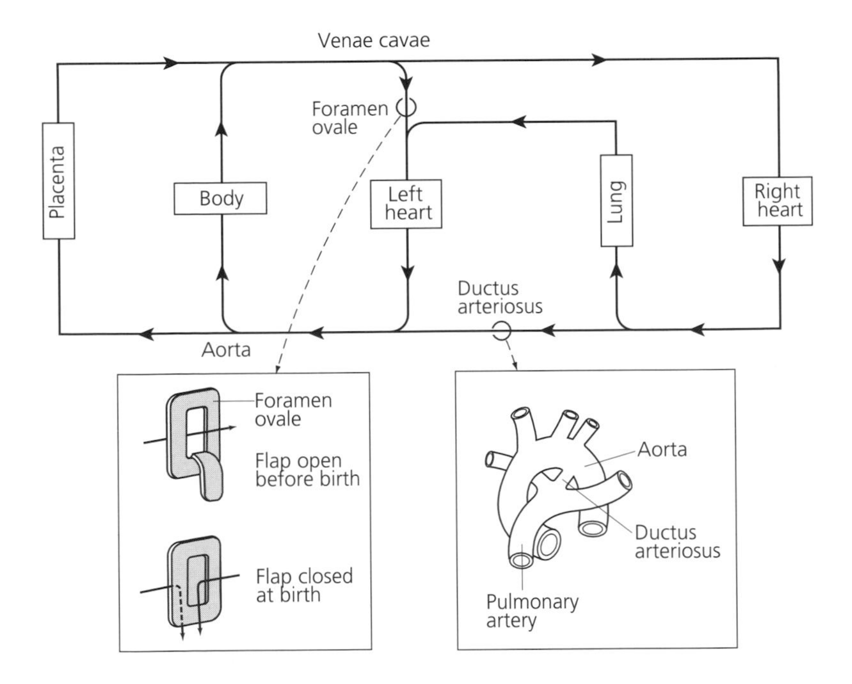

Figure 4.8 Fetal circulation.

4.9 Physiological and psychological effects of pregnancy

The physiological changes taking place in the pregnant woman are more dramatic than at any time in her life when she is in good health. Pregnancy is very demanding, both physically and emotionally. The woman gains weight, and as the pregnancy advances and the uterus grows, the abdominal muscles are stretched so that bending and getting up become more difficult. At the same time, she must prepare herself for the psychological demand of childbearing, as well as the physical stresses of labour and motherhood. Many women, however, feel very healthy and happy during their pregnancies, and feel few effects of the dramatic changes in their bodies.

Early indications of pregnancy include absence of menstruation, tender breasts and morning sickness. The latter symptom can be controlled by drugs, but in the wake of the thalidomide tragedy in the early 1960s (in which abnormal development of the limbs resulted in babies being born in some cases with deformed arms and legs after their mothers had taken the drug thalidomide for morning sickness) many women are understandably reluctant to take any drugs at all during pregnancy. However, morning sickness can be very debilitating, and hospitalisation may be necessary in severe cases.

The hormone changes brought about by the placenta are described in section 4.4, but there are also other general endocrine and metabolic changes, including increased thyroid and adrenal activity bringing about a general rise in metabolic rate. Cardiac output increases by about 30% in early pregnancy, and then stabilises. Blood pressure is normally slightly lower in normal pregnancy. Blood volume increases by about 30%, and plasma volume and red blood cell count both by about 45%. Ventilation rate increases due to an increased partial pressure of carbon dioxide ($^{P}CO_2$) in the arteries. Kidney function is stimulated, and glomerular filtration rate increases by about 50%.

The placenta is recognised as foreign by the mother's body, so her immune system undergoes changes in order that she does not reject it. Oestrogen induces the production of glycoproteins which circulate in the blood and coat certain white blood cells, thus hampering their ability to attack invading substances. It is also thought that the presence of embryonic cells in the maternal venous circulation may stimulate the woman's immune response. The action of molecules on the surface of the developing placenta and which would normally be recognised as foreign by the maternal T cells may also be altered in some way, suppressing the immune rejection. This is an active area of research, and if further light could be shed on the way in which a pregnant woman's immune system responds to the developing fetus, more may be learned about the problem of organ rejection after transplantation.

4.10 Antenatal health care

Over the last 300 years in the developed countries the rate of death in childbirth has fallen steadily from around 25 in 1000 in the mid-sixteenth century to less than one in 10 000 today. Of course it is difficult to say to what extent these earlier deaths were due purely to the pregnancy or birth, but it is estimated that around half of those deaths which occur around the time of birth are 'true maternal deaths'. Perinatal deaths are those of the baby between 28 weeks of gestation and one week after birth. At present this figure is around 10 per 1000 births, and shows a similar rate of decline over recent history to that of maternal death in childbirth. These statistics are directly related to the standard of antenatal care which is now available.

Antenatal clinics

Most women in this country, on finding themselves with a wanted pregnancy, attend an antenatal clinic. Here they will be examined by a doctor, specially qualified in **obstetrics**, the branch of medicine concerned with pregnancy and childbirth. This doctor's aim will be to look for any abnormalities which may affect the well-being of either the mother or the fetus. The doctor will assess the maturity of the fetus by finding out how certain the woman is of the date of her last menstrual period, whether her periods were regular and whether she used oral contraceptives prior to her last period. The woman will also have a blood test to check the level of her haemoglobin; to ascertain her blood group (ABO and rhesus); to see if she has any infections (including HIV in some circumstances); and to check for rubella antibodies.

At each visit, the midwife (a nurse specially trained in antenatal and perinatal care) or another nurse will monitor the woman's weight, and she will be asked for a sample of urine to test for protein and glucose to check that her metabolic balance is normal. Her blood pressure will be taken, and signs of swelling, e.g. around the ankles, which might indicate fluid retention will be noted. She will have her abdomen examined by the doctor or midwife, who will check the size of her uterus (figure 4.9), the heartbeat of the fetus and the position in which it is lying. Towards the end of the pregnancy, the presentation of the fetus and the engagement of its head at the brim of the mother's pelvis will be checked to ensure the fetus is in the head-down vertex position from which normal vaginal delivery begins (chapter 5).

Breech presentation – when after 32 week's gestation the fetus is lying with its head up and bottom towards the brim of the pelvis – represents a risk for the fetus. This is because the body is born first and the head may not be able to follow, by which time it is too late to opt for Caesarean delivery. The obstetrician makes a very careful examination of

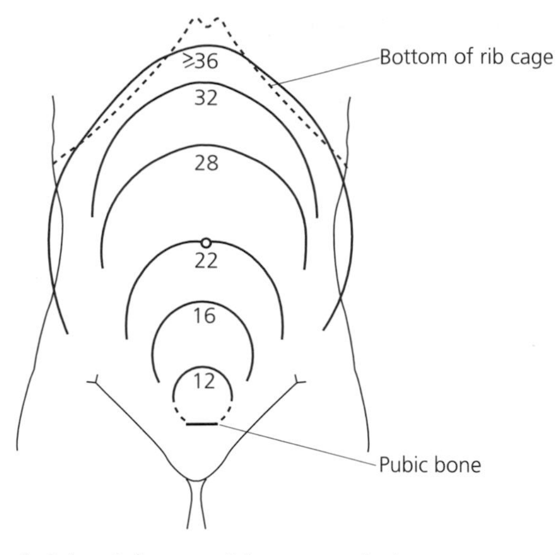

Figure 4.9 Increasing height of the top of the uterus during pregnancy (figures are weeks of gestation).

the pelvis when a breech presentation is discovered, and on the basis of this decides whether to try to turn the baby into the conventional vertex position, to do a Caesarean section, or to allow the labour to proceed without turning the baby.

Throughout her antenatal care the woman will be advised on such matters as diet (well balanced, plenty of protein, calcium and vitamins); smoking (avoid completely); alcohol consumption (limited consumption depending upon the stage of pregnancy); work (all right in moderation up to 32 weeks depending on the type of work and previous history of miscarriage); rest (needed towards the end of pregnancy); exercise (gentle throughout); sex (normal relations all right throughout, unless there is a previous history of miscarriage); and financial benefits. Advice during pregnancy changes continuously as more research evidence becomes available. For example, women are currently advised to avoid unpasturised milk and its products; not to handle pets too closely; and to avoid risk of contracting rubella. Most antenatal clinics organise parentcraft classes, which both parents are encouraged to attend. At these the parents-to-be are told about what is happening biologically, what to expect during labour, and how to look after the new baby.

Miscarriage

It is difficult to estimate the number of conceptions which do not implant, but it is thought to be relatively high. It is known that about one in seven of all diagnosed pregnancies abort in the first trimester, and that nearly all of these are abnormal. These events are called **spontaneous abortions**. Obstetricians do not become unduly worried about a woman's chance of carrying a fetus beyond this stage until she has suffered three spontaneous abortions. **Threatened abortion** is when bleeding from the vagina occurs during pregnancy. About three-quarters of women who experience this do not lose the fetus, but the rest abort and nothing much can be done to prevent this.

Beyond the first trimester, the causes of abortion include low position of the placenta in the uterus, and incompetence of the cervix, which dilates too early so that the fetus with its membranes cannot be retained. This can be remedied by stitching the cervix until the time of delivery. Although it can begin at any time, a major cause of death of full-term fetuses around the time of birth is placental failure. Women who may be at risk of this are monitored closely throughout their pregnancies. Placental failure may occur for a number of reasons, including poor health of the mother, bad diet, smoking, infections of the uterus and other diseases not associated with the pregnancy. Alternatively, the pregnancy may have gone past the due date, or more than one baby may be expected. Various tests, based on measuring hormone levels, can be done to detect placental failure. If nothing is done, the condition leads to retarded growth of the fetus in early pregnancy, and possible death or injury to the fetus in later pregnancy.

Complications of pregnancy

Obviously, this is a vast subject on which a very important branch of medicine is based, and antenatal care aims to identify and treat complications of pregnancy, concentrating, in order of priority, on the health of the mother and then the fetus. Various groups of women are at particular risk of complications in pregnancy. Those over 35 who are having their first baby may possibly have experienced difficulty in conceiving and, because of their age, the chance of doing so again is less than that of a younger woman. These older women are monitored closely: they are often not allowed to go past their due date for delivery of the baby, and they may be delivered in a specialist unit. If any problems occur during labour they are speedily delivered by Caesarean section (chapter 5). Women in their late thirties are at increased risk of chromosomal abnormalities, particularly trisomies (an extra chromosome 21, due to non-disjunction). This subject is dealt with in detail in *Heredity and human diversity* by Stephen Tomkins (CUP, 1989). The risk of producing a Down's syndrome child is about 1:1000 at age 25; about 1:100 at age 38; and 1:50 at age 45. Most clinics offer special tests (see later in this section) for this condition to women over 35. It should be noted that the advice and treatment that obstetricians give to pregnant women varies according to current research and their own training and experience.

Another high-risk group for complications in pregnancy are those women having their fifth or subsequent baby. Often the fetus will be unstable in the uterus, will be late to engage at the end of pregnancy, and the labour may be very rapid, with concomitant risk to the baby. With these possible problems in mind, the obstetric team will keep a particularly careful eye on this group.

Overweight women are also at risk, because of the possibility of high blood pressure and the difficulty in assessing the development of the baby by abdominal examination. Again, these women are monitored carefully during the antenatal period, and encouraged to stick to a diet which will minimise the amount of weight gained during the pregnancy.

It is common during pregnancy or delivery for a small number of fetal blood cells to come into contact with the maternal circulation. If the mother is rhesus-negative and is carrying a rhesus-positive fetus, she may be sensitised to produce anti-rhesus antibodies. In subsequent pregnancies these antibodies can cross the placenta and attack the red cells of the fetus. This in turn affects the liver of the fetus, causing it to be born with jaundice. It may even die before birth. A preventative substance called anti-rhesus immunoglobulin may be given to a rhesus-negative woman within 48 hours of delivery, and this disables any rhesus-positive cells in her circulation, preventing the stimulation of antibody production.

Detecting fetal abnormalities

Part of the initial examination of a woman in the early stages of pregnancy includes a discussion of her family history. Couples at risk of having an abnormal fetus include those with a family history of inheritable disorders such as sickle cell anaemia or PKU (this topic is discussed more fully in *Heredity and human diversity* by Stephen Tomkins (CUP, 1989), those who have had an abnormal child or fetus previously, and those women in their late thirties who are at increased risk of chromosomal abnormalities such as Down's syndrome which arise during oogenesis. Any couple in these categories will ideally be offered genetic counselling, which informs them about the nature of possible problems which may occur, what tests can be done and what are the implications of the outcome of the tests.

Embryo biopsy

A biopsy is a test in which living cells or tissue is examined. Embryo biopsy is not yet freely available, even to the most at-risk couples, and can only be done as part of an ***in vitro* fertilisation** (IVF) procedure. Because of the totipotency of the cells of the early embryo, a single blastomere may be removed by passing a micropipette through the zona pellucida without affecting its development (see section 4.2). This cell may then be examined for chromosomal defects (figure 4.10a). Using recombinant DNA technology, the presence of particular genes which have now been characterised, e.g. for Duchenne muscular dystrophy or cystic fibrosis, can also be detected. At the chromosomal level, the sex of the embryo can be determined, and this information may be valuable in the case of as yet unidentified genes that are known to be sex-linked. These tests may also be done on a blastocyst which has had a small incision made in its zona pellucida through which a small number of cells protrude (figure 4.10b). These are then removed and tested. If either the early embryo or the blastocysts are found to be normal, they can then be replaced in the woman's uterus in the usual way for an embryo transfer (chapter 8). The chance of the pregnancy succeeding is then about the same as for any other IVF procedure.

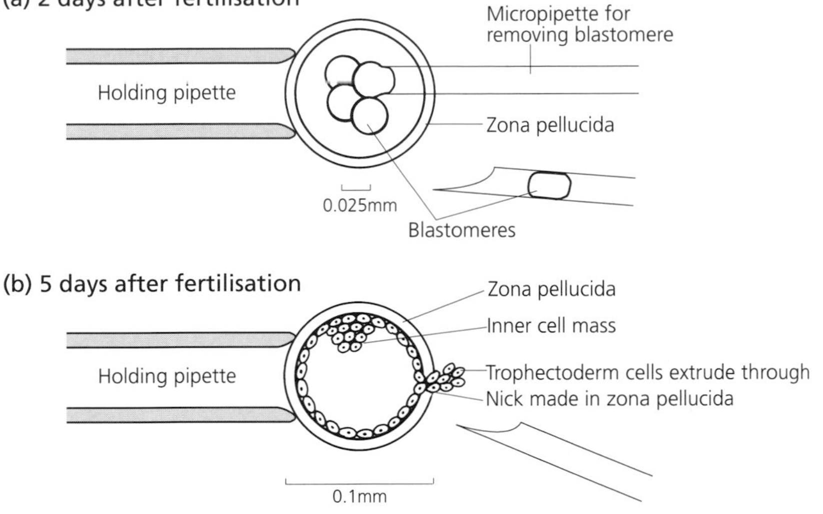

Figure 4.10 Methods for obtaining early embryonic cells for biopsy.

Ultrasound scanning

This is a widely used technique, and is routinely done in antenatal clinics for early confirmation of fetal heartbeat and for estimating fetal size as the pregnancy advances. It is also very useful for the detection and diagnosis of gross abnormalities. Ultrasound technology has advanced considerably in recent years, and it is now possible to make detailed measurements of various fetal body functions such as breathing movements and flow velocity in the blood vessels. A new technique called transvaginal colour Doppler ultrasonography can reveal details of structures within the body of the fetus.

Amniocentesis

In this test a hollow needle is passed through the walls of the woman's abdomen and uterus, and a sample of amniotic fluid is withdrawn. This contains cells from the fetus which, after culture, can be tested for chromosomal and genetic abnormalities. It cannot be done before 16 weeks of gestation timed from the date of the last menstrual period (14 weeks after fertilisation), and results are not usually available before the twentieth week of pregnancy. If the outcome of the test indicates a severe abnormality, the woman may be offered a termination of pregnancy. There is a slight risk of the procedure of amniocentesis itself bringing about a miscarriage.

Chorionic villus sampling

The advantage of this test over amniocentesis is that it can be carried out earlier in pregnancy. A needle is inserted via the cervix and guided by ultrasound scanning towards the placenta. Here, a sample of the fetal villus tissue is collected. Similar tests to those done following amniocentesis may then be carried out. Because there are usually more cells in the sample than the number obtained from amniocentesis, results are generally more quickly available because the cells do not have to be cultured prior to testing .

Fetoscopy

The fetus can be visualised directly by insertion of a fibre optic fetoscope. This may be done in addition to amniocentesis, or instead of it. In the latter case samples of tissue may be taken directly from the fetus or umbilical cord. As with amniocentesis, there is a small but significant risk of spontaneous abortion.

Maternal blood sampling

Again, this is a routine procedure carried out in antenatal clinics. Neural tube disorders of the fetus can be indicated at an early stage by an elevated level of a specific protein in the woman's blood, **alpha-fetoprotein**.

Future developments in antenatal medicine

Medical researchers are working on new techniques in embryo and fetal diagnosis, and these include use of nuclear magnetic resonance imaging, which can be used to build up serial images of the body. Potentially, this can give a direct view of the fetus without damaging it, and the use of various marker molecules will, it is hoped, give more detailed information about fetal physiological processes.

Already, techniques of antenatal surgery have been used to correct fetal heart, brain and bladder defects. These procedures involve opening the uterus and operating on the fetus directly. There is a high degree of risk to both the woman and the fetus, and so this type of operation is only attempted when there is a high probability that the fetus will otherwise die. Each operation of this type yields new information, but it is still uncommon and only undertaken in teaching hospitals associated with research into surgical techniques.

Cytology techniques involving the use of monoclonal antibodies are being employed to detect cells from the chorionic villi which break off and enter the maternal circulation. Once identified, these cells can be cultured and then subjected to chromosomal and genetic testing.

4.11 Termination of pregnancy

Abortion is the term used to describe the departure from the uterus of a fetus before it is able to survive independently. This may occur naturally, in which case it is usually termed spontaneous abortion or miscarriage (see section 4.10), or through medical intervention, when it is called **therapeutic abortion** or termination. The legal and ethical aspects of this subject are discussed in chapter 9.

The earlier in a pregnancy a termination is carried out, the safer it is. The procedure is carried out in an operating theatre under general anaesthetic. Most gynaecologists only perform late abortions in extreme circumstances. Counselling is invariably offered to women undergoing therapeutic termination of pregnancy.

The antiprogesterone compound **RU486**, also called mifepristone, the 'abortion' pill or the 'morning after' pill (see also chapter 6), brings about complete abortion. It has a **contraceptive** action if taken before implantation, or an **abortifacient** action if taken post-implantation. Administration of this drug is always carried out under medical supervision, and the after-effects are comparable with surgical termination. It is becoming an increasingly popular method of ending an unwanted pregnancy, as it does not involve anaesthetic or surgical intervention. This has particularly important implications for the developing world, where it is estimated that of the 125 000 women who die every year as a result of abortions, 95% are due to problems of basic hygiene.

4.12 Embryo research

Any research project which involves the use of live human embryos must, under the Human Fertilisation and Embryology Act 1990, be licensed. Careful consideration is given by the HFE Authority to any application for such a licence, and licences are only granted if it is decided that it is necessary or desirable:

- to promote advances in the treatment of infertility;
- to increase knowledge about the causes of congenital disease and to develop methods for detecting genetically or chromosomally abnormal embryos before implantation;
- to advance knowledge about the causes of miscarriage;
- to develop more effective methods of contraception.

All current techniques of gamete and embryo manipulation, such as IVF and embryo biopsy, were developed after many years of painstaking research. Such work would not have been possible without the help of people who allowed their embryos to be used in such work. At present the majority of embryos used in research come from IVF treatment. Under the law, a maximum of three embryos can be replaced in a woman's uterus during any one treatment cycle. A successful oocyte collection and *in vitro* fertilisation can result in more than three embryos developing. These 'extra' embryos can be either frozen for replacement in a subsequent cycle, or donated to research. Typical research projects which are currently underway include:

- measurement of metabolic activity in pre-implantation embryos;
- pre-implantation diagnosis of sickle-cell anaemia and haemophilia;
- investigation of the role of growth factors in embryogenesis;
- investigation of the effects of co-culture with endometrial cells on the viability of embryos;
- study of embryo cryopreservation;
- investigation on the effects of fluctuations in pH and temperature on the viability of embryos.

Basic scientific research such as this underpins all advances in our understanding of a particular subject, and it is essential that it is carried out meticulously and rigorously so that the results are reproducible and therefore enable meaningful interpretation. It is on this kind of work that techniques for treatment are founded.

FIVE

Birth and lactation

5.1 The perinatal period

Few areas of medical care have seen so many technological advances in recent years as that of the **perinatal** period (around the time of birth). This is particularly true of the techniques available to the late fetus and newborn baby. Childbirth is a natural function, but if it is left to run its natural course, it may not have a successful outcome for the mother or the baby. The obstetric team regard their job of helping a couple at the time of the birth of their baby as a privilege, and their aim is to make it as natural and happy an experience as possible. However, the safety of the mother and baby is of overriding importance, and if anything goes wrong during the birth they may have to intervene and manage the birth in a way which detracts from the parents' wishes.

Many couples favour natural childbirth, including home deliveries, with little or no medical intervention. However, there is no doubt that the delivery suite in a hospital is in a position to act very quickly if an emergency arises.

5.2 Stages of normal labour

Labour is an appropriately named body function – it is very hard work. The staff of the antenatal clinic prepare the prospective parents for the birth of their child by organising classes at which the process of labour is explained and their own plans for the delivery are discussed. The mother-to-be will have already been booked in for the time of the birth, and the parents will be told to telephone the maternity unit as soon as labour has reached a particular stage.

Towards the end of the pregnancy, the woman may experience 'practice contractions' of the uterus, which may be mistaken for the onset of labour. On the other hand, some women do not feel the contractions at all in the first stage of labour. However, labour usually begins with regular uterine contractions which become increasingly intense and more frequent, the loss of the amniotic fluid from the vagina ('breaking of the waters') or, in some cases, loss of mucus and a small amount of blood from the vagina. The whole process of labour tends to take longer for the first baby, and becomes progressively quicker with subsequent births.

The first stage of labour is characterised by strong contractions of the uterus during which the cervix becomes increasingly **dilated**. For a woman having her first baby this stage may last 8–10 hours, and for women having subsequent babies 6–8 hours, although these times are highly variable. During this time the cervix dilates at an average rate of about 1 cm per hour, and the baby's head descends into the mother's pelvis and becomes stabilised there (figure 5.1). As this is a protracted stage of labour, the woman is encouraged to move about until the contractions become so frequent that she wishes to get into a position ready for the actual delivery. During this stage she does not push with her abdominal muscles to help the contractions, but will have been given advice and training on how to breathe to ease the process and to pace herself through the contractions.

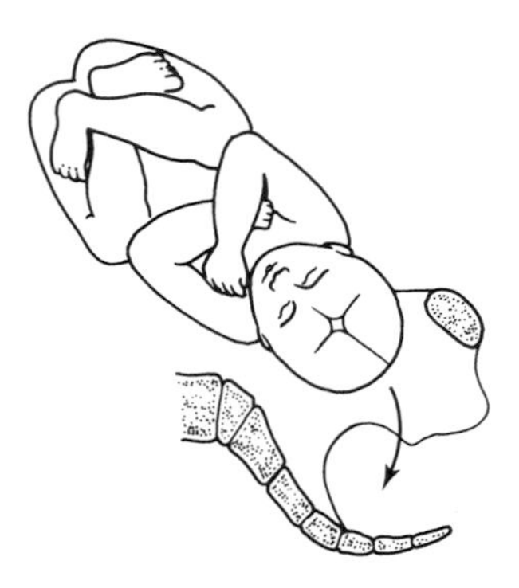

(a) Head enters the true pelvis

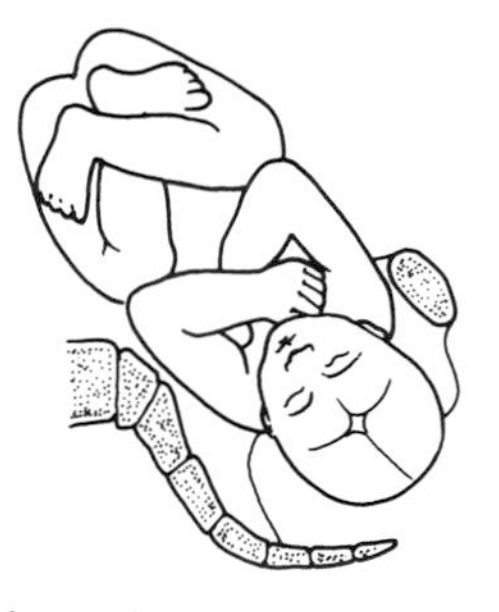

(b) Head descends to pelvic floor

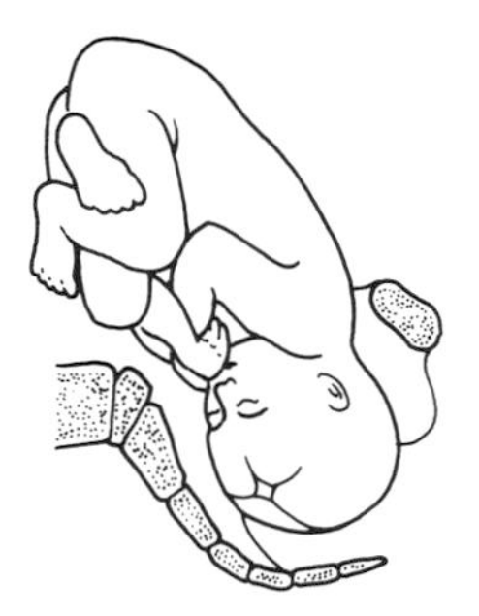

(c) Internal rotation, back of head comes to lie beneath pubic bone

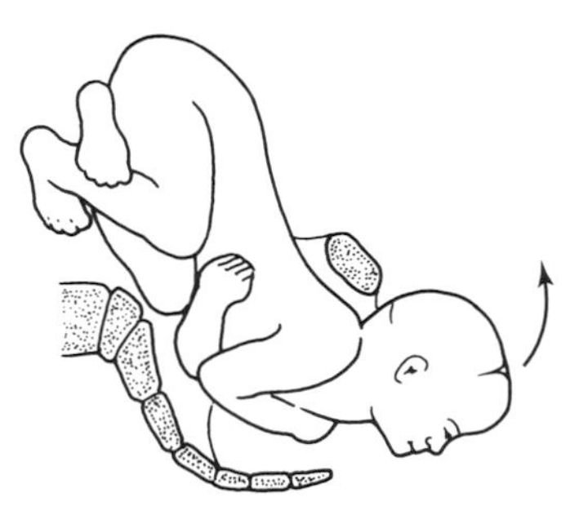

(d) Further descent, head is born

Figure 5.1 Stages of normal delivery.

The second stage of labour is much quicker, usually lasting 10–60 minutes, depending upon whether it is the woman's first labour. By now the contractions are almost continuous, and the woman is encouraged by the midwife to squeeze with her abdominal muscles at each contraction to help push the baby out. In fact the urge to do so is usually overwhelming. The vagina dilates and, if the presentation is normal, this allows the baby's head to descend further, turning as it does so. The back of the baby's head then reaches underneath the pubic arch of the front of the mother's pelvis, and further contractions bring the head out of her body.

The final stage is very rapid, consisting of a few contractions which bring about the birth of the rest of the baby's body, followed after a while by the delivery of the placenta and membrane. As the baby's shoulders are born, an injection of **oxytocin** is usually given to the mother. This causes a very strong contraction of the uterus which separates the placenta from the lining. The midwife pushes down on the mother's lower abdomen as the baby is born. The placenta is carefully inspected to make sure it is complete, as remnants left inside the uterus cause bleeding and possible infection.

The baby is usually put to the mother's breast for a short time before the umbilical cord is clamped and cut about 5 cm from the baby's abdomen. The baby is wrapped, its airways checked, and then it is weighed. Immediately after birth, major physiological changes must take place to bring about the adaptations necessary for life outside the uterus. The circulation must be redirected because the lungs rather than the placenta are now the site of gaseous exchange (figure 4.8). This process is described in more detail in section 5.4. The alimentary canal begins to secrete digestive enzymes, as the main dietary constituent, milk, is rich in lipid and protein, and is ingested and absorbed in the gut. During fetal life glucose will have been supplied to the bloodstream of the fetus via the placenta. The baby also has to acquire control of the regulation of its body temperature. Newborn babies are unable to shiver to generate body heat, and so this is done by the metabolism of **brown adipose tissue** which is deposited around the vital organs. The oxidation of this reserve generates sufficient heat to keep the baby's body temperature up, although it is very important to ensure that a newborn baby is kept in a comfortably warm room, away from draughts, and is well wrapped up.

5.3 Medical intervention

Pain relief

Because birth is a hazardous time for both mother and baby, medical intervention is not uncommon. Frequently, for example, as labour is usually an extremely painful process, pain-relieving drugs are given. Often these analgesics are injected in the early stages of labour, or given by inhalation at a later stage. **Epidural anaesthetic**, injected directly into the spine, is often given during labour, and it gives total pain relief without loss of consciousness. However, the woman cannot feel anything at all below the level of the epidural, and cannot therefore push with the contractions as well as she could if she were aware of their onset. Decisions about administration of pain relief are made carefully to enable the woman to assist with the birth and to remain alert so as not to miss this very significant event in her life.

Fetal distress

For the baby, each contraction stops the blood flow in the spaces between the chorionic villi, thus reducing the oxygen supply. Under normal circumstances, and when the labour is not too protracted, the baby is able to compensate physiologically for this. However, if the oxygen level to the baby is reduced too much, a condition called **hypoxia** may ensue, with potentially disastrous results. Therefore, the obstetric team monitor the well-being of the baby carefully throughout labour. The heart rate is measured either externally by stethoscope through the wall of the mother's abdomen or by ultrasound. If the baby's head is already in sight at the dilating cervix, heart rate can be monitored by electrocardiography, taken by attaching a probe to its scalp. The rate should remain between 120 and 160 beats per minute.

Another sign of fetal distress is the appearance of **meconium** in the amniotic fluid. Meconium is a dark faecal substance which accumulates in the fetal gut during the third trimester. Under conditions of low oxygen (hypoxia), the gut may contract and the anal sphincter relax. In order to check this, the amniotic sac is ruptured once labour is fully underway, if it has not already done so naturally.

If the fetal heart rate slows, or meconium is seen, a sample of the fetal blood may be tested to see if it is becoming too acidic. Fetal blood pH falls below the normal level of 7.2 if hypoxia occurs, because of the accumulation of carbon dioxide and lactate. When these levels are observed, the team waste no time in delivering the baby immediately, because under these circumstances it could die very quickly.

Forceps and Caesarean deliveries

If the delivery has to be hastened, for example if the mother is too exhausted by the labour to be able to help push the baby out, or if the second stage of labour is delayed, the obstetric team may use forceps. These have a curved blade that fits into the mother's pelvis and around the fetal head. The doctor can thus apply traction, or pull and rotate the baby to bring it into the world. The technique is only used when the head is engaged, and at least at the level of the mother's pelvic brim.

The name for the surgical procedure in which the mother's abdomen and uterus are opened to remove the fetus and membranes did not actually come from the method of Julius Caesar's birth as commonly supposed. Roman law held that if a woman was dying during late pregnancy or birth, the fetus had to be cut from her body. It has been practised throughout history, but only in the last century was there any realistic chance of both mother and baby surviving it. In those days the incision was vertical, but now the most common method is the **lower segment Caesarean section (LSCS)**, in which the incision is transverse low in the abdomen. The LSCS involves less blood loss, and post-operative recovery is quicker. It is also less disfiguring, as the scar is just above the line of the pubic hair.

Sometimes the doctor will decide to perform a Caesarean section before the woman goes into labour. This is called an **elective Caesarean**, and is usually done the week before the baby is due, at 39 weeks' gestation. There are various reasons for this, including most commonly:

- placental inadequacy leading to retarded growth of the fetus;
- previous gynaecological surgery including other Caesarean sections;
- malpresentation of the fetus;
- diabetes;
- high blood pressure.

More commonly, though, Caesarean sections are performed as emergency procedures during labour. The woman may be given a general anaesthetic, in which case her first sight of the baby is when it is wrapped, clean and weighed, often sleeping in its cot or its father's arms. It is possible to have a Caesarean section by epidural anaesthetic, and this has the advantage that the woman is conscious during the birth and can see and hold the baby as soon as it is delivered, just as she would be able to following a normal vaginal delivery. Recovery from Caesarean section is usually quite quick, and the woman is encouraged to get up and about and to look after and feed her baby almost as quickly as if she had had a normal birth.

5.4 Physiological changes

Very soon after birth, the circulation through the umbilical cord ceases. This creates a resistance against the blood coming from the left side of the baby's heart, and this is diverted to the pulmonary arteries which expand in response to the newborn's efforts to breathe. Increased circulation in the pulmonary arteries causes a lowering of the blood pressure in the aorta and this in turn brings about the contraction of the ductus arteriosus. The pressure changes which occur as the umbilical arteries close down initiate closure of the foramen ovale. The fetal cardiovascular structures are not fully closed down for several days. More information on changes in fetal circulation following birth is given in section 4.8.

5.5 Lactation

The mammary gland

The lactating breast, or **mammary gland**, is perhaps the most obvious distinguishing feature of the mammals – the vertebrate group to which human beings belong. It is believed that mammary glands evolved from specialised sweat glands in the chest skin of very early forerunners of mammals, possibly as early as 250 million years ago. Humans, in common with other primates, only have two mammary glands but other mammals can have up to 25, as is the case in some marsupials. Mammary tissue consists of numerous ducts ending in alveoli that consist of secretory cells. It is well supplied with blood vessels and surrounded by contractile epithelial cells. The alveoli synthesise and secrete the milk, and this is released along the ducts leading to the nipple by the action of the contractile cells. The mammary tissue itself is surrounded by fatty (adipose) tissue (figure 5.2).

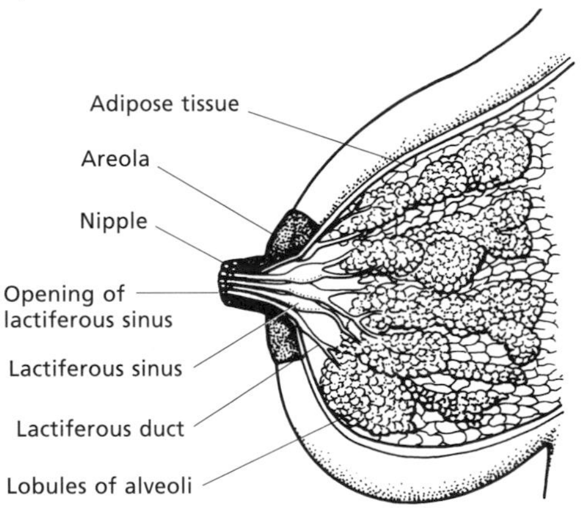

Figure 5.2 Anatomy of the breast.

Milk production

Lactation is the term used for the synthesis and secretion of milk from the mammary glands. During adolescence, and increasingly during pregnancy, the breasts develop under the influence of ovarian hormones. These hormones stimulate growth of the duct system (oestrogen) and growth of the secretory tissues (progesterone). During the first trimester of pregnancy, the alveoli start to secrete a milk-like liquid, but proper milk production (**lactogenesis**) develops during the third trimester under the influence of the hormone prolactin from the anterior lobe of the pituitary gland. Prolactin is prevented from exerting its full lactogenic effect by the high levels of oestrogen in the maternal circulation during pregnancy, but as soon as birth has occurred the level of oestrogen drops, which allows the onset of lactation. Suckling of the nipple by the baby stimulates the flow of oxytocin from the posterior pituitary, and this stimulates the contractile cells in the mammary glands to squeeze the alveoli, causing 'let-down' of the milk. In the first week after **parturition** (birth), a substance high in antibodies, known as **colostrum**, is produced. Although it looks thin and watery colostrum is in fact very nutritious.

Over the following weeks the milk flow is maintained by stimulation from the hormones prolactin and oxytocin, secretion of which is in turn stimulated by the suckling of the baby. The more the baby stimulates the breast in this way, the more milk is produced. At the height of lactation, as much as 800 ml can be obtained from a woman's breasts at one time. However, the breast is not a milk store, like a bottle of milk. In fact, only a small amount is stored immediately prior to feeding the baby, and this accounts for the uncomfortable feeling of fullness at this time. The rest of the milk consumed during a feed is actually produced whilst the baby is sucking.

Milk is a complete food for the baby, not only providing the required balance of nutrients, but also conferring immunity from disease in the early stages of lactation as a result of the antibodies it contains (the antibody content of breast milk decreases over time). Breast milk is also readily available, at the right temperature, sterile and free! For all these reasons, women are encouraged to breast feed their babies if at all possible, at least for a while.

Milk yields about 3.44 kJ cm^{-3} of energy, and when a woman is breast feeding a growing baby with no supplementary milk, this places a considerable demand upon her body. For this reason, it is most important that she eats a good balanced diet with plenty of energy-rich and calcium-containing foods. She also needs to drink more fluids than usual to replace that lost in the milk. It is interesting to note that in cases of malnutrition of the woman, whilst she is able to maintain lactation the quality of her milk is not diminished until the malnutrition is severe.

Prolactin levels fall progressively during lactation, but with sufficient stimulation by suckling prolactin production can be maintained for as long as four years. When the baby is suckling frequently and stimulating a good supply of milk, the woman may not ovulate or have a menstrual period. This is termed **lactational amenorrhoea**, and is associated with a high level of prolactin in the blood. There is evidence to suggest that the more episodes of breast feeding that occur per day, the higher the levels of prolactin. So as the baby begins to sleep for longer periods during the night, and consequently does not demand food then, so the likelihood that ovulation will occur increases. In addition the sudden increase in circulating LH (the LH surge) which heralds ovulation usually commences at night, therefore it is thought that the frequency of breast feeding at night may be a significant factor in determining the return of fertility. If the prolactin level falls sufficiently, ovarian function may be re-established, and a woman may ovulate as early as five weeks after delivery.

Our early female ancestors, who possibly had no knowledge about how to limit the number of children they had, or perhaps had no desire to do so, probably experienced a reproductive life of alternate pregnancy and lactation. Repeated menstrual periods would have meant infertility, and although these are now regarded as normal by modern women, they are not a feature of the naturally evolved human life cycle.

5.6 The puerperal period

Careful preparation for parenthood, particularly motherhood, can help people considerably come to terms with the great changes in their mental, physical and social states brought about by having a baby. The **puerperium** is the period taken for recovery from the effects of pregnancy and delivery and on average is about two weeks, providing there have not been any complications. Over this period the uterus gradually returns to the normal non-pregnant size, a process called **involution**, during which the extra tissue of the uterine wall is reabsorbed and a substance known as **lochia** is discharged. The lochia contains mainly blood and endometrial cells, and gradually, over a period of 3–6 weeks, it becomes clear and finally stops.

During the hours and days immediately after the birth, the parents, and especially the mother, usually experience a deep psychological affinity towards their baby. This is known as **bonding** and it is of the most profound importance for the well-being of the infant, and satisfies a basic biological drive in the parents. The mother's physiological response to the birth supports this. Post-natal care is aimed at promoting bonding by encouraging the parents, from the moment of the baby's birth, to spend as much time as possible holding and caring for it.

S I X

Contraception

6.1 World-wide issues of population growth

It has been estimated that the population around the first century BCE was around 250 million. Disease, war and famine were the main reasons why it took another 1730 years for it to reach 1000 million. However, it has been suggested that throughout history people have practised birth control by a variety of methods, such as bringing about abortion, prolonged lactation, deliberate celibacy, *coitus interruptus* (withdrawal of the penis from the vagina before ejaculation), various alternatives to 'natural' sexual intercourse, and even infanticide. Figure 6.1 shows what has happened to the world population since the year AD 1.

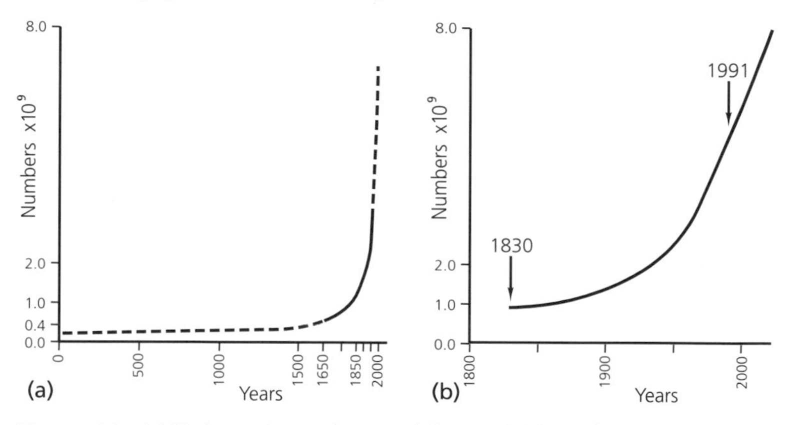

Figure 6.1 *(a)* Estimated population of the world from the year AD1 to AD2000 *(b)* World population since 1830.

Sources adapted from: (a) After Dorn, 1966; *(b)* Controlling reproduction, Hutchinson, Chapman and Hall.)

One glance at these graphs will show the extent to which the pressure on resources exerted by the human population has increased over the last 200 years. Many factors have helped to bring this about, principally increased survival to reproductive age and a lowering of the age of puberty due to improved nutrition and health in many parts of the world. As a result of this rapid increase in population, events of human history during the past 200 years have centred largely around a more intensive exploitation of the biotic and abiotic environment and the resulting territorial disputes.

At the time of writing, the world population is around 5400 million. By the year 2000, another 800 million people will have been added to that figure, and 95% of that increase will be in Africa, India and South East Asia. The population of the Northern Hemisphere has, by and large, stabilised, and the problems that beset it environmentally and economically are those contributed by an ageing population and high levels of gaseous and aqueous pollution per head of the population. The developing countries of the Southern Hemisphere, on the other hand, have a rapidly enlarging population with an emerging problem of habitat destruction and terrestrial pollution.

This picture is now complicated by the appearance and effect over the last 10 years of the human immunodeficiency virus (HIV). It is estimated that by the year 2000, around 3 million people will have died as a result of AIDS-related illnesses, and around another 120 million will have been infected by HIV. In some parts of Africa around 1:40 people are HIV-positive, and the death rate due to complication with other life-threatening diseases such as tuberculosis is high enough already to have affected the economic development of these areas.

The control of human fertility is now possible for both men and women who wish to exercise it, and who live in circumstances where it is possible to do so. This means that in certain societies education about, and availability of, contraception enables people to be sexually active, yet to separate recreation from procreation. This has caused a profound shift in political, religious and legal thinking in those countries where people are in a position to choose this option. The moral and ethical aspects of this are discussed in chapter 9.

6.2 Mechanisms of available methods

The methods of contraception which are currently available were developed in the main during the 1950s and 1960s. Although a great deal is now known about reproductive function, the 'perfect' form of contraception still remains elusive. The following descriptions of contemporary contraceptive techniques are organised under headings corresponding to their site of action.

Gamete production

Regulation of gamete production is currently only available for women. Progress on the development of a 'male pill' is discussed in subsequent sections of this chapter, but when the relative numbers of gametes produced by men and women are compared, it is easier to control an event that occurs only once every 28 days. This fact, probably combined with the vested interest of women for whom conception results in pregnancy and all the ensuing consequences, led to the development in the 1960s of oral hormonal contraceptive methods. The effect of this on society has been enormous: for the first time women were empowered to choose whether and when to have children, and in turn this enabled them both to marry and have a family and to pursue a career. In the UK, only 10% of married mothers worked in 1900, but by 1976 this proportion had risen to 50%.

There are two main types of contraceptive pill, and these vary in the relative amounts of oestrogen and progesterone that they contain. The combined pill contains synthetic forms of both oestrogen and progesterone, and these hormones are taken daily for the first 21 days of the ovarian cycle. There is then a break during which time bleeding occurs. The hormones contained in the pills exert their effect on the hypothalamus, inhibiting the secretion of GnRH, thus confusing the natural pituitary–ovarian relationship. The natural feedback system to the anterior and posterior pituitary is interrupted, with the result that the LH surge does not occur. This effectively suppresses ovulation. In addition to this action, the combined pill also changes the receptivity of the cervical mucus which would normally aid sperm travel into the uterus, and it also prevents the endometrium preparing for implantation. This all makes it an extremely effective fertility control. Originally a course contained 21 pills which all had the same composition, but biphasic and triphasic pills are now available which more closely mimic the action of the natural hormone profile. This form of contraception can also be administered by once-a-month intramuscular injection.

The second type of contraceptive pill, which contains progesterone only and has a different contraceptive action, is discussed under 'Mucus receptivity'.

Gamete passage within the body

Sterilisation

People who have decided that their families are complete may choose to undergo surgery to ensure permanent sterility. Female sterilisation involves severing and tying the Fallopian tubes to ensure that sperm and oocytes cannot meet. There is a very low failure rate, but occasionally the tubes reunite and fertility returns. Women who may change their minds after having had the operation cannot guarantee that they will again be able to conceive naturally.

Male sterilisation is also known as **vasectomy**. The vasa deferentia are severed and tied. Sperm cannot leave in the ejaculate, and are reabsorbed in the epididymis. Sperm production continues, however, and potency is unaffected by vasectomy. Microsurgical operations to reverse vasectomy in men who change their minds are really only successful in the hands of experts as, like the Fallopian tubes, the ducts of the vasa deferentia are extremely narrow and delicate.

Natural methods

Contraception which does not employ any permanent or artificial intervention is known as the natural method, and it is based on accurate prediction of ovulation and avoidance of sexual contact at that time. It is the only acceptable form of family planning for many people (see chapter 9), and while there is at least a 2% failure rate (2 pregnancies per 100 couples practising the method per year), there are no harmful side-effects, and some people find that it benefits their sexual relationship. The advocated procedure is known as the **sympto-thermal method**, and uses a combination of careful calendar recording, daily monitoring of body temperature, and checking of the cervical mucus changes which occur around the time of ovulation. With good motivation, it can be as effective a method of birth control as artificial contraceptives.

Coitus interruptus, otherwise known as 'withdrawal', cannot be regarded as a form of contraception, as small numbers of sperm are discharged prior to the main ejaculation.

Barrier methods

Contraceptive techniques aimed at preventing the sperm from reaching the oocyte inside the woman's body are called barrier methods. The condom is made of very thin rubber and is placed over the erect penis. It is coated with a lubricant containing a spermicidal agent and, when used properly, is an effective form of contraception. Condoms are readily available and easy to use, but they must be put on before there is any contact between the penis and vagina. They can slip off, and should be held in place as the penis is withdrawn after intercourse. Some men find that condoms reduce sensitivity during intercourse, and lovemaking may be interrupted by having to put them on. The use of condoms has been encouraged among the sexually active single population because of the protection they provide against sexually transmitted infections such as HIV.

Women wishing to take responsibility not only for contraception but also for prevention of transmission of disease have the option of using the so-called female condom, e.g. Femidom and Femshield. This recently developed device is a disposable sheath which is stronger and lighter than the latex rubber used in a condom. When in place, it lines the vagina.

The diaphragm is a circular rubber cap with a springy elastic rim which fits over the cervix. It is used in conjunction with spermicide cream which is applied around the rim before the cap is inserted. It has to be put in before intercourse, and taken out not less than six hours afterwards.

Diaphragms are checked for fitting by family planning experts, and can be washed and reused. They interfere less than condoms or Femidoms with the spontaneity of lovemaking, and are relatively easy to apply, but they do have a failure rate of about 2%. A less effective variation on this method is the contraceptive sponge, which is impregnated with spermicide and placed high in the vagina. It must remain in place for the same period as the diaphragm, after which it is discarded. This makes it considerably more expensive than a diaphragm.

Mucus receptivity

An early hurdle for sperm in their journey through the female tract towards the site of fertilisation is the cervical mucus. At the time of ovulation it is naturally receptive to the sperm, and aids their journey through the cervix and into the uterus. Any system which upsets this receptivity has contraceptive potential.

Another type of oral contraceptive, the progesterone-only pill, is sometimes also called the 'mini-pill' because of its low dose. This means that the hypothalamus is not completely suppressed, and ovulation may occur. Its contraceptive action is on the epithelial cells of the cervix, which it causes to produce thickened mucus. This forms an effective barrier against sperm. The action of the cells of the endometrium is also affected, making implantation unlikely. It is possible to have this form of contraception administered by once-a-month injection, and recently implants have been developed which are effective for up to five years.

Implantation

Implantation is prevented by the action of several types of contraceptive device, including the progesterone-only pill and its injected or implanted counterpart, the 'morning after' pill (see section 4.11), and the intrauterine device.

RU486 - the 'morning after' pill, an anti-progesterone steroid

This synthetic compound was developed in France in the late 1970s but it was not until 1988 that it was first used in the termination of pregnancy. It has remained controversial ever since. RU486, also called mifepristone, works at two levels: pre-implantation, as a form of emergency birth control (the 'morning after' pill), and post-implantation when it can bring about abortion (chapter 4). If taken within three days of unprotected intercourse, a single dose of RU486 will ensure that pregnancy does not result.

Intrauterine device (IUD)

Otherwise known as the coil or loop, IUDs are small, copper-coated plastic appliances of various designs which are inserted by a doctor through the cervix and into the cavity of the uterus. The IUD remains in position until it is removed by pulling thin plastic strings which remain attached to it. It has about a 2% failure rate, but it is considered suitable for women who have had children and who, for various reasons, cannot take oral contraceptives. It works by disrupting the endometrium by abrasion, and so preventing implantation. For this reason, some women who use it report heavier or more painful periods. It also puts the woman at greater risk of infections of the reproductive tract. Another serious disadvantage of the IUD is that implantation may occur in a position other than the uterus. This is known as ectopic pregnancy and is discussed in section 4.3.

6.3 Recent advances in contraception

No method of contraception is without its disadvantages. Furthermore, the advantages and disadvantages of any method will often be viewed differently by the people using them and by the people providing them. Taken together these facts add up to a considerable challenge for reproductive biology. The following have been suggested as attributes of the 'perfect' form of contraception. According to these criteria, contraception must:

- have a specific action and not interfere with or harm any other body process;
- be 100% efficient at preventing conception;
- be usable continuously, possibly for many years;
- be immediately reversible;
- not be too expensive to produce or administer;
- not affect libido or sexual satisfaction;
- be easy to use;
- have a margin of safety to take into account human failure.

Refinement of existing methods

Steroid-releasing vaginal rings

This technique has been under development for many years, and consists of a ring of silicone rubber placed high in the vagina and left for a period of up to three weeks. Steroid hormones contained in the ring are released slowly but continuously. This has the advantage over the pill of giving a low and uniform dosage, thus avoiding the 'spike' of synthetic steroid in the blood, not seen in the natural ovarian cycle, which results from the oral contraceptive.

Steroid contraceptives for men

Although for many years it has been technically possible to halt spermatogenesis, it has, until recently, been considered unacceptable. Now, in some parts of the world, attitudes have changed such that men will consider alternating responsibility for contraception with their partners.

The method under development consists of a weekly injection of steroid hormone (androgen) which results in azoospermia (absence of spermatozoa in the semen) after 2–3 months in about 75% of men. Sperm count, quality and motility are restored when the injections are discontinued for a similar period.

RU486

This drug has been used for some time for the medical termination of pregnancy, based on a 600 mg dose followed by administration of a prostaglandin. It has recently been suggested that RU486 could be used as a once-a-month contraceptive pill if taken at the time of ovulation. It has also been observed that if RU486 is taken during the follicular phase of the cycle, oocyte development is disrupted and ovulation inhibited. It has therefore been proposed that a daily low dose of RU486 for three weeks followed by a daily dose of progesterone would provide effective contraception. This is still under investigation.

Other hormonal approaches

GnRH (gonadotrophin releasing hormone) analogues

GnRH is produced by neurones in the hypothalamus and acts on the pituitary gland, stimulating the pulsatile release of LH. Many artificial GnRH mimics (analogues) have been produced, some of which have promoting (agonist) properties whilst others are inhibiting (antagonist) in action. Both can be used to block the release of LH and to some extent FSH. This in turn suppresses folliculogenesis in the ovary. These substances therefore have contraceptive potential for both men and women, but their continuous use, as would be necessary to give protection, may put women at risk from breast cancer and osteoporosis (bone demineralisation). GnRH analogues are used routinely in IVF treatment (chapter 8), but have to be used in conjunction with hormone replacement therapy (HRT) if administered over an extended period. In men, research is underway using GnRH antagonists to produce reversible azoospermia without loss of libido, but their use as a contraceptive is still some considerable way off.

Gonadotrophin antagonists

Work is in progress to produce recombinant FSH antagonists which will mimic FSH action, but will not interact with receptors in the ovary or testis.

Maternal recognition of pregnancy

The corpus luteum (CL) has a vital function in providing hormonal support for pregnancy until the placenta takes over (chapter 4). The CL itself is maintained by human chorionic gonadotrophin (hCG), which is secreted by the early embryo. If hCG is absent, menstruation follows. It would seem reasonable, then, to try to develop an hCG antagonist. However, the receptor to which hCG binds appears to be identical to that for LH, so blocking it would result in the disruption of oestrogen and progesterone production in the luteal phase of the ovarian cycle. The way forward which is currently being explored is subtly to change the conformation of the hCG molecule itself by attaching an extra peptide to it. This has the effect of preventing antigen–ligand recognition, and thus neutralising the effect of hCG. This technique is known as immunoneutralisation.

Some researchers find this advance difficult to accept, because effectively an hCG vaccine would act as a block to maternal recognition of pregnancy rather than preventing pregnancy itself. Contraception methods, to be acceptable, must be reversible, and a vaccination has the connotation of permanence. On the other hand, as an alternative to sterilisation, lifelong vaccination might have some attraction. For the future, it is not impossible to foresee that one vaccination, given at puberty, might confer life-long contraceptive protection, with the immunological block being lifted selectively when required.

Interception of sperm-oocyte interaction

One of the causes of infertility is the presence of antibodies produced by both women and men to spermatozoa. When these antibodies are present, the sperm are seen to clump together or to stick to the cervical mucus. The events of fertilisation are complex, and so offer a variety of opportunities for interception at the immunological level. However, two main events of fertilisation seem to present the best way forward at the present time:

1. *Sperm–zona binding*
   A sperm receptor protein on the zona pellucida, ZP3, is well known, and work is now in progress to modify the ZP3 such that it cannot bind with the sperm or initiate the acrosome reaction.

2. *Sperm–oocyte surface membrane binding*
   This approach is to disrupt the proteins which cause binding of the sperm membrane following acrosome reaction with the oocyte surface membrane.

Molecular delivery systems

Molecular biology is providing tools for many branches of science and medicine. Its techniques are being used in basic research, the diagnosis of disease, and in the manipulation of various life processes, including reproduction. Finding ways to manipulate the reproductive system at its molecular level is an enormous challenge.

A molecular delivery system is one in which fertility can be regulated by administering genetically modified or synthetically produced substitutes for natural hormones or releasing-factors. Recombinant DNA technology is advancing at an unprecedented rate, and it is now possible to produce the complex peptide sequences which make up such hormones in great quantity and at relatively low cost. The challenge for reproductive biology is to find a route of administration, because taking these substances orally subjects them to the action of digestive enzymes. Currently, the preferred route is nasal spray or intramuscular injection.

The development of gene insertion techniques is proceeding apace, and the possibility now exists for certain body tissues to be engineered to produce such proteins as CnRH analogues, FSH-binding proteins or oocyte surface membrane receptor proteins.

SEVEN

Human infertility: background and causes

In view of the increasing problems of world over-population, it may be surprising to learn that human fertility is relatively poor. Compared, for example, with rodents, reproduction in people is very inefficient, and the fact that there are so many of us is due to improved survival rather than reproductive success. Even in couples who have already conceived (i.e. are of proven fertility) the average monthly chance of the woman getting pregnant is only 20–25%, and just by chance about 10% of these fertile couples will fail to conceive during their first year of trying. It is only after at least this period of time, during which their family doctor will have probably explained to them carefully how to maximise their chance of conceiving, that they are referred to a specialist fertility clinic.

The complete inability to conceive a child – infertility – is very rare. This would only be the case, for example, if the woman had completely blocked Fallopian tubes or premature menopause, or if the man had a complete lack of sperm. Absolute infertility in both partners who are of reproductive age means that there are no treatment options open to them, and their only means of having a family would be to adopt or foster children. The anxiety of wondering whether they will be acceptable as foster parents or adopters can be as stressful as the treatment for infertility.

Much medical and scientific research is focused upon causes of and treatments for subfertility, which may be defined as difficulty in conceiving naturally for reasons affecting the man, the woman or both partners. The causes of subfertility may be treatable by various clinical interventions. It was due in part to public concern about research into subfertility treatments that the Human Fertilisation and Embryology Act was passed in 1990. A government 'watchdog' called the Human Fertilisation and Embryology Authority (HFEA) was set up to regulate and monitor treatment centres, and to ensure that research using human embryos is carried out in a responsible way.

The HFEA does this by means of a licensing system, which covers the following areas of activity:

- any fertilisation treatment which involves the use of donated eggs or sperm, or embryos created outside the body;
- the storage of gametes and embryos;
- research on human embryos.

7.1 Problems of subfertility

Studies have shown that as many as one in six couples seek specialist help at some time in their reproductive lives because of difficulty in getting pregnant, and that a similar number may be unable to have a baby, but do not seek help. This means that subfertility is a fairly common problem. Most people have a basic instinct to have children, and when they learn that this may not be possible, they may experience a form of sorrow similar to the grief of bereavement. During the course of their lives, most people can expect that others close to them will die, but that in time the sadness which results from it will diminish. The inability to have children, however, is a lifelong affliction – people continue to grieve for the children they never had. It has been termed 'genetic death', and those whom it affects suffer very badly and need help.

Treatment for subfertility has only been available in the latter part of this century. In times gone past, it was very much a taboo subject, and childless women were said to be 'barren' and were often pitied or even reviled. It is an interesting social phenomenon that infertility was almost invariably assumed to be the woman's problem, possibly because male infertility was often disguised by infidelity, the woman's children having been fathered by other men. Nowadays, however, issues such as involuntary childlessness are more openly and frequently discussed, and with increasing scientific and medical understanding of the causes, considerably more can be done to help the subfertile.

One of the most important aspects of subfertility treatment is counselling. To help them to come to terms with their problems, subfertile couples need information, and to talk to someone who is both informed and sympathetic about the implications of childlessness and/or fertility treatment. Most young couples trying to start a family do not expect that they may be unable to; subfertility comes as a shock, and of course it affects not only the subfertile person, but his or her partner as well. It has been seen that the worries and anger that the subfertile couple may experience are often focused around their treatment. This means that to be effective in the long term (whatever the outcome of the treatment) and to help prevent other problems such as their relationship breaking down, the counselling of the couple must be closely linked to the treatment. Many counsellors are now specialising in infertility counselling, which means that patients can have the benefit of talking to someone who is not only independent of the medical and scientific staff who are involved in their treatment, but also has the time and expertise to listen to and help them to deal with their anxieties.

7.2 Conditions needed for conception

The basic requirements for conception are shown in figures 7.1 and 7.2. Failure of any of these will result in subfertility or infertility. These requirements are numbered more or less in the order in which they occur, but some are much more likely to fail than others. Table 7.1 shows the relative frequency of the various common causes of infertility.

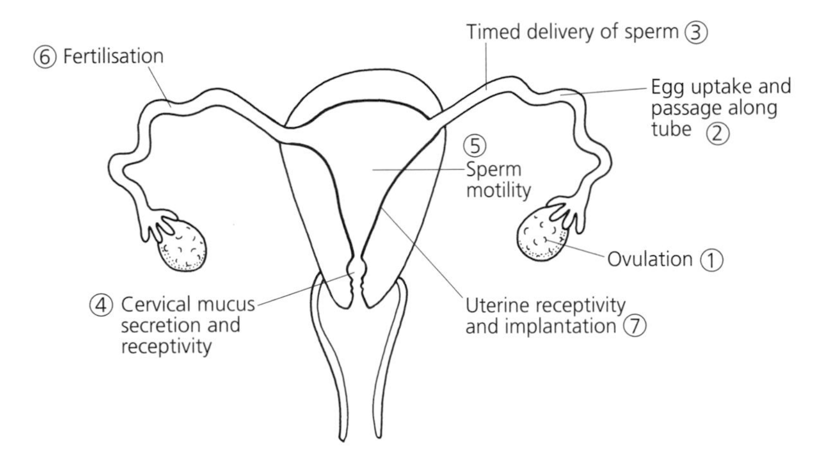

Figure 7.1 Conditions needed for conception in women.

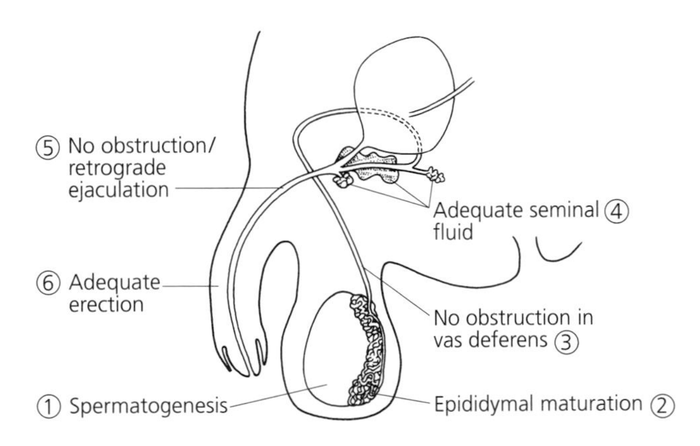

Figure 7.2 Conditions needed for ejaculation of viable sperm in men.

Table 7.1 Relative frequency of the various causes of infertility

Cause	*% Couples*	*Cause*	*% Couples*
Ovulatory failure	21	Sperm defect/disorder	24
Tubal damage	14	Other male factor	2
Endometriosis	6	Coital failure	6
Mucus defect/disorder	3	Unexplained	28

The total comes to more than 100% because some couples have more than a single cause.

Source: Hull M.G.R. *et al.* 1985. British Medical Journal **291**: 1693.

7.3 Preliminary diagnosis of infertility and subfertility

The extent of medical investigation of couples who fail to conceive varies considerably. This section covers most of the preliminary tests which may be carried out by GPs before referral to a specialist clinic. Both partners will be given a physical examination, and the doctor may ask them about any previous problems, e.g. irregular periods or urinary tract infections. The doctor will also need to know how often and when they have intercourse. The male partner may be asked to provide a semen sample

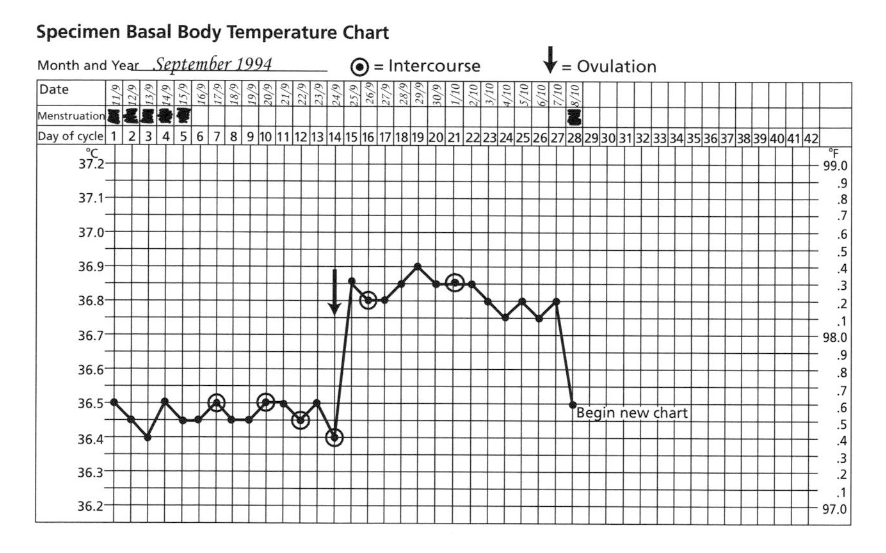

Figure 7.3 Specimen basal body temperature chart.

which will show if the sperm count and motility are normal. The woman may be asked to record her temperature at the same time each day on a **basal body temperature (BBT) chart** (figure 7.3) to see if the changes are regular and normal (chapter 2). The doctor may decide to check that ovulation is occurring by taking blood samples at about day 21 of a 28–day cycle (day 1 being the first day of the woman's menstrual period). The amount of progesterone present in her blood gives an indication of how well her corpus luteum is functioning. Since luteal function depends closely on follicular function, a good luteal phase is only seen when a good follicular phase has occurred. Both temperature recording and progesterone testing have to be done over several monthly cycles because even in women whose cycles seem regular there is still some variation.

Armed with the information from these initial tests, the couple may then be referred to a fertility clinic. Here, a specialist – usually a consultant gynaecologist with a particular interest in infertility problems – will do more detailed tests. These will help him or her to make a diagnosis and to decide upon the best form of treatment.

7.4 Causes of infertility

Ovulation failure

In view of the complex hormonal interactions that bring about ovulation (chapter 2) it is not surprising that this is the greatest single cause of female infertility. Often an ultrasound examination of the ovaries at mid-cycle as an additional check on ovulation will be done. Failure to ovulate is associated with **amenorrhoea** (complete lack of periods) or **oligomenorrhoea** (infrequent and irregular periods). Sometimes, the BBT chart or progesterone tests at around day 21 may indicate that although the woman's periods are regular she has problems with ovulation, or the luteal phase is inadequate. There are five main causes of ovulation failure:

1. feedback problems of oestrogen to the hypothalamus;
2. problems with production of FSH and LH by the pituitary, and hence lack of oestrogen production by the ovaries. Amenorrhoea is the usual symptom;
3. failure of the ovaries to respond to FSH – this situation is found at the menopause;
4. the ovaries may have many small cysts (polycystic ovary disease, PCOD). FSH and oestrogen levels are normal, but LH is raised in this condition;
5. abnormally high level of prolactin, and reduced FSH and oestrogen (hyperprolactinaemia).

The fertility specialist can determine the cause of ovulation failure from the results of a hormone assay, in which the relative amounts of LH, FSH, prolactin, oestrogen, testosterone and thyroid hormone are measured.

Tubal obstruction

Any impediment to the passage of the egg from the ruptured follicle to the site of fertilisation in the Fallopian tubes will cause infertility. The specialist will therefore need to establish that the Fallopian tubes are open and healthy (patent). The most commonly used procedure is a laparoscopy and dye test. Under a general anaesthetic, the woman's uterus, tubes and ovaries are examined directly through a special telescope (laparoscope) which is inserted into her abdomen via a small incision near the navel. Coloured dye is passed into the uterus via the cervix, and any blockage which impedes the progress of the dye along the tubes can be seen through the laparoscope. Laparoscopy can also establish whether there is any reason why the oocyte may not pass from the ovary to the Fallopian tube, or if a condition called **endometriosis** (see below) is present. Less frequently, the specialist may assess the condition of the inside of the uterus and tubes using a test which involves laparoscopic examination followed by X-rays taken after injection of a radiopaque dye into the uterus and tubes.

The Fallopian tubes are not merely connections between the ovaries and the uterus, but delicate and intricate structures with a subtle and active physiology which creates a precise microenvironment for the passage of gametes, fertilisation and early embryo development. Damage to the tubes and pelvic adhesions are usually due to inflammation in the pelvic area. This may be the result of infections such as peritonitis or appendicitis, or infections from the uterus, perhaps as a result of a complication in a previous pregnancy. This damage may take the form of tissue adhering to the surface of the ovary, or thickening of the delicate membranes at the end of the Fallopian tube so that the egg cannot pass into the tube. Alternatively, the lining of the tube itself may be damaged so that the egg cannot pass along towards the uterus. Such damage may also affect the tubal microenvironment, making it unsuitable for the final preparation of the gametes for fertilisation. In some cases, the tube is actually blocked or, in the case of surgical sterilisation, severed (chapter 6).

Endometriosis

In this condition, small pieces of endometrium (lining of the uterus) are found outside the uterus. Commonly they are found on the ovaries and the ligaments that support the uterus. When the woman has her menstrual period, these patches of endometrium bleed as well, and because there is no normal escape route for the blood it accumulates, causing pain and infertility. Endometriosis can interfere with the motility of the tubes, or give rise to obstruction of the ovaries and fimbria.

Mucus defects

The cervical mucus serves an important function in receiving, storing and selecting sperm cells. Increased secretion of the mucus at mid-cycle, just before ovulation, is under the control of oestrogen, the level of which is high at this time in the cycle. Under the influence of oestrogen, the mucus becomes thinner and more elastic due to the mucoprotein strands which it contains aligning themselves in parallel.

The woman may be asked to come to the fertility clinic for an examination of her cervical mucus at the time of ovulation when a **post-coital test** (PCT) may be carried out. This involves taking a sample of the cervical mucus 6 to 12 hours after the couple have had intercourse, and examining it under the microscope. If more than about 30 motile sperm cells are present per low-power field of view, then it can be concluded that the man is producing sufficient sperm, that they are being ejaculated in the right place and that they can survive in the cervical mucus.

Mucus penetration tests may also be carried out. Here, a drop of mucus collected from the woman's cervix at day 13 or 14 of her cycle is put onto a warmed microscope slide, and then a drop of the man's semen is placed on the slide next to it. A coverslip is placed over the preparation, and the progress of the sperm is observed. Similarly, a two-way penetration test may be done, in which the couple's mucus and semen are tested against donor mucus and semen which are known to be compatible. The interpretation of all these tests is complex, however if the sperm fail to progress through the mucus or become immobilised, there is a possibility that the mucus may contain anti-sperm antibodies which disable the sperm. Blood tests can confirm that such antibodies are present.

Sperm defects

Human semen is notorious for its variability, and sperm disorders are the single most common cause of infertility. A routine semen analysis involves measuring the volume of the ejaculate, counting the sperm cells in a counting chamber of known volume, estimating the percentage of sperm cells that are forwardly motile, and estimating the percentage of sperm cells which are of normal structure. Because of the length of the spermatogenic cycle (64 days), two semen analyses are usually carried out, often a few weeks apart. The range of normality of semen is very wide, and the sample of an individual man can vary according to his lifestyle and state of health. However, it is unlikely that he will be fertile unless his ejaculate contains more than 30 million normal motile sperm cells per millilitre of semen.

Sperm count can be greatly affected by over-heating of the testes, which need to be at a lower temperature than the rest of the body for normal functioning. It is for this reason that they lie outside the body cavity, within the scrotum. Hot baths, tight clothing and sedentary jobs can all affect sperm count, as can very frequent ejaculation, tiredness, stress, smoking and excess alcohol.

While 'normal' semen always contains some abnormal sperm, some men can have a normal sperm count, but have a high proportion of abnormal sperm cells (figure 7.4). Sometimes examination of the semen sample will reveal clumping (agglutination) of the sperm, preventing free motility. This is due to autoantibodies on the sperm membranes themselves, which causes the sperm to stick together. This can also be seen in mucus tests. A rare condition is sometimes seen in which the semen contains good numbers of normal sperm, but although they are alive they are completely immotile. This is because they lack a component of the flagellar apparatus of the tail.

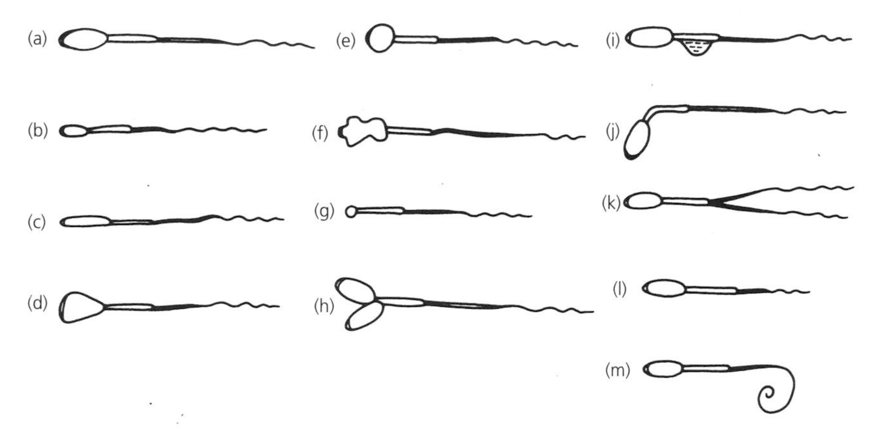

Figure 7.4 Sperm polymorphism. In the ejaculate of a normally fertile man, about 40% of the spermatozoa are 'abnormal' in structure. It is not known why this is so but it is not seen in closely related species such as the great apes. (*a*) Normal. (*b*) micro. (*c*) tapering. (*d*) pyriform, (*e*) round, (*f*) amorphous, (*g*) pin, (*h*) bicephalous, (*i*) cytoplasmic droplet, (*j*) bent, (*k*) biflagellate, (*l*) short tailed. (*m*) coiled tailed.

Other male factors

Absence of sperm, or a low count, can sometimes be attributed to a partial blockage of the vas deferens or the epididymis. Other causes of reduced fertility include the presence of varicose veins around the testis and vas deferens. The increased amount of blood brought near to the testis by these veins can affect sperm production because of the associated rise in temperature. A similar condition in which fluid collects around the testes is another common cause of male subfertility. Occasionally, the prostate gland becomes chronically infected and this can inhibit the fertilising ability of the sperm. Hormone disorders are less common causes of infertility in men than in women, however imbalance of FSH, LH, testosterone or prolactin can all be contributory factors. Rarely, retrograde ejaculation, in which semen is ejaculated backwards into the bladder, is a problem for some men, notably those with spinal injuries.

Unexplained infertility

Frustratingly, the most frequent cause of infertility is unexplained. Even after extensive investigation of both partners, no apparent problem is discovered. Obviously this means that no specific treatment can be prescribed. However for couples with unexplained infertility of less than three years, most are within the normal range of fertility and are likely to conceive naturally within a further two years. If they have been infertile for more than three years, it is unlikely that the woman will get pregnant, and the best hope for having a child that is genetically theirs is to have assisted reproduction treatment (chapter 8).

7.5 Options for the infertile couple

If complete infertility is diagnosed in both partners, they must either come to terms with their childlessness as best they can, with or without counselling, or opt for adoption or fostering. There are relatively few healthy babies available for adoption in this country nowadays, mainly because of improved contraception and more social acceptance of single parenthood. There has been an increase in recent years in the number of couples adopting babies from comparatively poor, and over-populated or war-torn countries. The procedures involved in this are often far from straightforward, and many people question the morality of removing children from their native country even if they might have a better upbringing. Many children who require fostering have a physical, mental and/or emotional handicap, and caring for them may require a rather different commitment than bringing up children who do not suffer from these types of problems. However, although fostering is often a short-term arrangement and parting can be very painful for both the child and the foster parents, it has enormous rewards and many people have reaped great happiness and fulfilment from it. Some people may decide that neither adoption nor fostering is an option for them, and they must then learn to get on with their lives without having a family of their own.

There are more options open to the couple if only one partner is infertile, and even more if the specialist decides that the infertility is in fact subfertility which is therefore treatable. These alternatives will be discussed in chapter 8.

E I G H T

Human subfertility: options and treatment

Problems of subfertility affect both sexes approximately equally. Chapter 7 dealt with initial investigation and diagnosis of both female and male patients, and this chapter will discuss the options and treatment available to the subfertile couple in specialist centres and clinics. Many of the procedures of reproductive technology are surrounded with considerable moral and ethical controversy, and the final chapter of this book, chapter 9, is devoted to a discussion of the main issues.

Relatively speaking, less is known about causes and few remedies are available for problems of male subfertility, though it is the focus of an increasing amount of research. The study of male reproduction is called **andrology**.

8.1 Problems of gamete production

Treatment options for poor sperm quality

Techniques for improvement of sperm quality

Although a few cases of **azoospermia** (total lack of sperm in the semen) have been treated successfully by administering FSH and/or hCG, very little can be done at present about either azoospermia or **asthenospermia** (poor sperm quality), and the couple is usually advised to consider donor insemination (see below). **Oligospermia** (very low sperm concentration) can often be helped by taking general measures to improve health, such as losing weight, stopping smoking, reducing alcohol intake and avoiding stress. Some doctors and medical scientists have tried using various hormones and other chemical substances such as arginine, zinc and vitamin E for treating low sperm count and/or poor sperm quality but with very limited success. Assisted reproduction remains the best hope for a man with such a diagnosis, and the various techniques (described in the following sections) are aimed at overcoming the natural barriers presented to the sperm in its journey to fertilise an oocyte, thus bringing the egg and sperm closer together.

Artificial insemination techniques

In **AIH** (artificial insemination by husband's – or partner's – sperm) a sample of the semen is first carefully prepared by centrifuging, taking off the supernatant and resuspending in a special medium. This procedure is often repeated, and the final concentration of sperm cells is accurately adjusted by suitable dilution to a particular volume. This sample is then loaded into a small syringe gun to which a thin tube (catheter) is attached, and the sperm cells, in their medium, are deposited at and around the top of the woman's vagina or cervix. **Intrauterine insemination** (IUI) is a variation on this, and is a procedure in which the catheter is passed through the woman's vagina and cervix, this putting the semen directly into the woman's uterus. The chance of successful conception is maximised by carefully monitoring the time of ovulation (chapter 2). When IUI is used, the volume of the sperm sample has to be about 50 microlitres, and this involves very careful preparation of the optimum number of sperm, all of which have to be of the highest quality.

Men for whom artificial insemination is a suitable treatment for subfertility include those who have a problem with normal sexual intercourse, perhaps caused by impotence, failure to ejaculate properly or spinal injury. Artificial insemination is less successful in men with oligospermia. IUI can also help couples in which the woman has identified subfertility such as producing cervical mucus which is hostile to the sperm. These methods are also employed in cases of unexplained infertility, though they are only occasionally successful. Most couples are counselled carefully beforehand and their hopes not raised too high. Artificial insemination techniques can also, of course, be used with donor sperm (see below).

Donor insemination (DI)

The method of depositing the sperm in the woman's reproductive tract is exactly the same as for AIH, but the sample used originates from an especially recruited, fertile donor. Donor sperm may also be used in other fertility treatments, such as IVF or GIFT. The couples for whom it offers hope of having a baby which is genetically the woman's include those where:

- the man has oligospermia or azoospermia;
- the man has asthenospermia (abnormal sperm);
- the man has had a vasectomy which is unreversed;
- the man has had surgery, radiotherapy or injury to his reproductive tract;
- the couple have incompatible blood groups;
- the man is a carrier of a serious genetic disease.

Some clinics will also treat women who have no male partner, but wish to have a baby without having a sexual relationship (chapter 9).

Treatment of the couple by DI is the culmination of a lot of commitment by many people. There are three main steps: recruitment and screening of donors; testing, freezing and preparing the sperm; and finally, insemination.

Sperm donors

Relatively few DI centres have their own donor clinic. This is because there must be a large population of potential donors to make it worthwhile. Towns and cities with a large student population often have donor clinics, but recruitment can also be from other sectors of the population, such as police, fire and ambulance services and also from men of proven fertility who have decided to undergo vasectomy. Those centres which do not recruit their own donors buy in frozen samples from larger centres.

Potential donors are first interviewed to find out their family and sexual history, so that possible chromosomal and serious sexually transmitted diseases may be identified. For example, someone who has had a close relative die of Huntington's chorea or who admits to a promiscuous lifestyle would probably be dissuaded from sperm donation at an early stage. Travelling expenses only are paid to donors, so they must be very committed because they may have to donate as often as twice per week for many months. They must be aged between 18 and 55, and be fit and healthy. Each donor is asked to produce a semen sample by masturbation. This is then analysed to assess its volume, density (sperm count), motility, morphology and presence of antibodies, and also to screen for infections such as hepatitis, syphilis and HIV. To be acceptable for storage, three samples, three months apart must have at least:

- sperm count greater than 60 million cm^{-3};
- ejaculate volume greater than 2 cm^3;
- 50% progressive motility;
- greater than 40% freeze–thaw survival;
- 50% normal sperm morphology;
- no antisperm antibodies;
- negative for viral or bacterial infections.

The donor's physical characteristics are recorded, and when a donor is selected prior to insemination, the choice is based on a match with the social father-to-be. The criteria for matching donors and recipients are usually: ethnicity; body build; hair and eye colour; height; blood group; and religion.

Donated sperm freezing and preparation

If a sperm sample meets all the criteria listed above, subsequent samples from a donor are prepared in the laboratory by dilution with a **cryoprotectant** (a protective agent which prevents damage to the cell membranes and contents during freezing) and is then drawn up into specially labelled tubes called straws. The straws are placed for about 10 minutes in the vapour of liquid nitrogen, during which time they cool rapidly to about minus 80°C. They are then put into the liquid nitrogen which is at minus 196°C where they can stay indefinitely. This process is known as **cryopreservation**. After six months of donation and storage, the man is screened again for infections such as HIV. If this proves negative, his samples can be released for use in treatment.

DI is most often performed by a specialist fertility nurse, who will have monitored the patient's menstrual cycle until ovulation. At this time, the nurse selects a donor straw, takes it out of the tank of liquid nitrogen and allows it to thaw for 10–15 minutes at room temperature. Having updated the storage records, the nurse puts a drop of the sample onto a microscope slide to check that there are motile sperm in it. For intravaginal or intracervical deposition, the nurse then loads the contents of the straw into the catheter for insemination. For IUI, the sample must first be washed and prepared as described above for a fresh sample.

Insemination with donor sperm is carried out using the same procedure as that described above for artificial insemination with fresh sperm from the woman's partner. The donor, although the genetic father of any child born as a result of DI treatment, has no legal obligation to the child, and has the right to remain anonymous. This is an important function in the UK of the Human Fertilisation and Embryology Authority, which keeps a record of all sperm (and oocyte) donors but maintains their confidentiality. If a person over 16 who knows or suspects that they were conceived as a result of DI (or oocyte donation) inquires of the HFEA about their genetic parentage, the Authority can tell them if they were born as a result of a donated gamete, and also whether they are related to anyone they may wish to marry. The actual identity of the donor however, is never disclosed. It is interesting to note that at present the legal limit for the number of pregnancies from any one donor is 10, and as soon as a donor reaches this figure his sperm cannot be used in further DI treatment and must be discarded. Obviously, therefore, the chance of one person born as a result of DI treatment finding out from the HFEA that the person whom they wish to marry was born as a result of gametes from the same donor is extremely remote.

Success rates for DI average around 10% for intracervical and up to 20% for IUI per cycle of treatment. Couples are usually advised that they should think of their treatment as a six or eight month course, rather than a series of individual treatments. It has been estimated that 50–60% of women undergoing a series of six treatment cycles become pregnant.

Treatment options for ovulation failure

In view of the complex hormonal interactions that bring about ovulation (chapter 2) it is not surprising that this is the greatest single cause of female infertility. Failure to ovulate (**anovulation**) is often, but not always, associated with amenorrhoea or oligomenorrhoea (absence of, or irregular, menstrual periods). Alternatively, menstrual cycles may appear completely normal but are anovulatory. Failure to ovulate can be caused by a failure of any part of the hypothalamic–pituitary–ovarian axis (figure 2.2).

Most women who experience ovulatory problems have adequate gonadotrophic hormones, and therefore production of ovarian hormones normally follows, but there is failure of the feedback loop between the hypothalamus, the pituitary and the ovaries. This condition is called hypothalamic pituitary dysfunction, and it is estimated to be the cause in 95% of women with anovulation.

In these cases, a substance called **clomiphene citrate** is used to treat the subfertility. This is an anti-oestrogen drug which mimics the action of oestrogen, tricking the brain into taking up the clomiphene in preference to oestrogen. The hypothalamus cannot monitor the level of the natural oestrogen in the woman's blood, and perceiving that this must be low stimulates the pituitary to secrete FSH. This in turn stimulates the ovary to produce more oestrogen, and also brings about maturation of Graafian follicles. In fact, women for whom clomiphene is the treatment of choice usually have normal levels of oestrogen. Patients normally start a course of clomiphene treatment with fairly low doses. Those receiving this type of treatment are monitored carefully by ultrasound scanning, often in conjunction with hormone testing, to check the number of follicles which mature. If the LH surge is adequate, ovulation of the oocyte brought to maturity by clomiphene will occur at the appropriate time.

For some women, even when on the maximum dosage of clomiphene ovulation does not occur. This is due to a deficient luteal phase, and administration of hCG, which mimics the action of LH, is required. In some cases a single injection of this is enough to bring about ovulation and a sufficiently sustained luteal phase to give an early embryo an increased time to implant, but often a second dose is necessary.

The output of GnRH can be disturbed by injury to the hypothalamus, presence of a tumour or, less directly, by effects from other parts of the brain as a result, for example, of over-exercising, emotional or mental disturbances and eating disorders. This particular problem can be treated with pulsatile GnRH administration, which mimics the action on the pituitary of natural GnRH, thus allowing the rest of the axis and its feedback systems to function.

Imbalance of the pituitary hormones can lead to anovulation, but this is less common. Excessive prolactin production may interfere with the normal secretion of FSH, and amenorrhoea or oligomenorrhoea may result. This condition is usually treated with **bromocriptine**, a drug that is very effective in restoring normal hormone levels; ovulation usually follows.

Oestrogen deficiency also causes ovulation failure. These patients often have poor FSH output, and the treatment involves boosting this with human menopausal gonadotrophin (hMG). The ovaries are very sensitive to this, and follicle development is always monitored carefully.

A common cause of ovulation failure is polycystic ovary disease (PCOD) (chapter 7) in which a large number of follicles begin development, but because the ovaries and other tissues produce abnormal amounts of the male hormone androgen, these follicles do not mature and never ovulate. The condition can be treated with clomiphene.

Oocyte donation

For a woman who is unable to produce oocytes at all but who has a functional uterus, donated oocytes offer the only hope of becoming pregnant with a child which is genetically that of her partner. The donated oocytes are placed in a specially prepared dish containing culture medium with the partner's sperm, and replaced in the usual way for a routine IVF treatment cycle (see section 8.2). For two important reasons, oocyte donation cannot be regarded as directly comparable to sperm donation. First, oocytes are in much shorter supply, as a donor must go through the procedure of ovarian stimulation and oocyte retrieval (described in section 8.2), and may yield at best only a few oocytes. Understandably, very few women are able to make this commitment. Secondly, freezing oocytes is still at the research stage as the delicate meiotic spindle apparatus may be irreversibly damaged by the freezing procedure, and the oocyte therefore unable to divide after it is thawed. This means that complicated management of recipients and donors is necessary if suitable oocytes are to be made available to women requiring this type of infertility treatment.

For these reasons, donated oocytes are in extremely short supply. They are needed not only for treatment of infertile women, but also for research on such projects as development of a protocol for oocyte cryopreservation; development of a contraceptive vaccine; and development of micromanipulation techniques. Potential donors must be between 18 and 35 years old if the oocytes are to be used in the treatment of others. They must also be offered counselling and given full information about the implications of their donation. In common with sperm donors, an oocyte donor is required to give full details of her medical history and of any inherited diseases in her family. She will also be tested for such infections as hepatitis B and HIV. Any child born as a result of treatment with her donated oocytes, although genetically her offspring, will not be legally hers and, as with sperm donors, the HFEA maintains a register of oocyte donors but does not disclose identifying information.

An alternative source of donated oocytes is from ovarian tissue, derived from live donors, cadavers or aborted female fetuses. In 1994 the Human Fertilisation and Embryology Authority published a consultation document on this issue, and wide-ranging public debate followed. As a result of this it was decided that the use of oocytes from cadavers or aborted fetuses should not be permitted in fertility treatment.

8.2 Problems of gamete interaction

Blockage of the male reproductive tract

Any blockage of the male reproductive tract which prevents the maturation and subsequent passage of sperm through the excurrent ducts, epididymis, vas deferens or sperm ducts will compromise fertility. This may have been caused by disease or carried out surgically as, for example, in vasectomy (chapter 6). More rarely, congenital abnormalities may result in blockage. The ducts of the male reproductive tract are extremely narrow, but developing microsurgery techniques can sometimes re-establish patency of these tubes with the result that sperm can again be present in the ejaculate.

If this procedure fails, the only recourse is to a new method called microepididymal sperm aspiration (MESA) in which a very few immature sperm are withdrawn, under anaesthetic, directly from the epididymal duct using a very fine needle. They are then used in assisted reproduction procedures (see below). MESA is not generally available, as there are only a few practitioners able to do it. The failure rate is very high. If the entire epididymis is blocked, it has now become possible to remove sperm still undergoing spermatogenesis from the testis and inject them directly into oocytes (see below). This procedure is still essentially experimental, although there have to date been a few live births resulting from it.

Blockage of the Fallopian tubes

Tubal obstruction in the female tract is a common cause of infertility, and can also be remedied by microsurgery. Providing the blockage is not too complicated, this technique is relatively successful. All microsurgery requires magnification of the field of operation with a microscope or magnifying spectacles worn by the surgeon. Smooth instruments, often made of glass, are used to avoid further damaging the delicate lining of the tubes. In an alternative to tubal surgery the Fallopian tubes are bypassed completely using assisted reproduction methods.

Assisted reproduction

This is a general term which means helping a couple to conceive by manipulating their own, or donor, gametes outside the body. The two main methods are artificial insemination (see section 8.1) and *in vitro* fertilisation (IVF) and its variations.

In vitro fertilisation (IVF)

In vitro means 'in glass', and it is either in a test tube, or, more usually, in a type of petri dish that the gametes of the man and the woman are mixed in the laboratory in such conditions that fertilisation *in vitro* may occur. Following fertilisation and early development, up to three embryos may, under the Human Fertilisation and Embryology Act 1990, be replaced in the woman's uterus. The success rate, as measured by the number of live births per treatment cycle, is increasing, but must still be regarded as poor (figure 8.1). The national average in 1990 was 13% per treatment cycle.

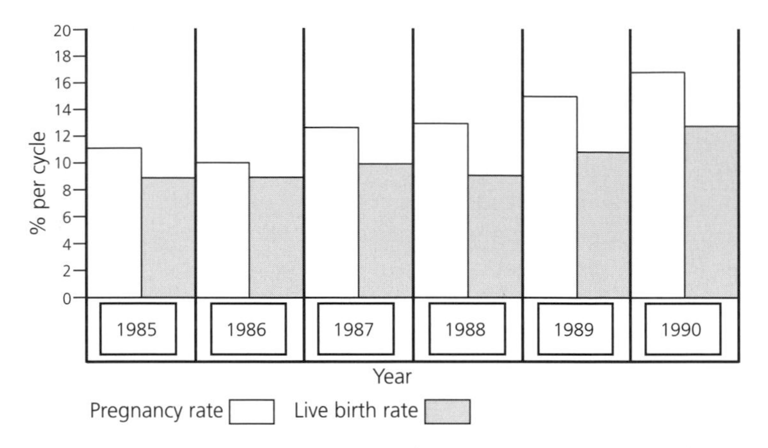

Figure 8.1 Results of nationwide IVF. These results, compiled by the Human Fertilisation and Embryology Authority from data sent to them by centres undertaking IVF treatment, shows the pregnancy and live birth rate figures per cycle of treatment. 1985–1990.

Source: Adapted from HFEA, 2nd Annual Report, 1993

However, in some treatment centres the live birth rate from IVF can reach as high as 30%. It should be borne in mind that this high rate is probably due in part to careful selection of patients – for example, younger women, those who have already had children and those with simple reasons for subfertility have a far greater chance of success with IVF.

Many types of couples with subfertility can be helped with IVF:

- women with blocked Fallopian tubes;
- men with oligospermia that is not so severe that there is no realistic chance of fertilisation;
- couples in whom the sperm and cervical mucus are incompatible;
- women with endometriosis but whose ovaries are still functional;
- couples who have more than one cause of subfertility;
- couples who have unexplained infertility;
- women who have complete ovarian failure and who are receiving donated eggs fertilised by their partner's sperm.

IVF is a complex and very demanding treatment for all who are involved in it. The main stages are:

1 *Stimulation of ovulation during the treatment cycle (sometimes called 'superovulation')*

The best chance of getting pregnant from IVF occurs when more than one embryo is replaced in the uterus. A regime of hormone administration is needed to mature more than one Graafian follicle in each cycle. Commonly, the pituitary is 'down regulated' by using a GnRH agonist which blocks the release of FSH and LH. This gives greater control over the ovaries, whose natural output of oestrogen is stopped. Shortly after this is started, injections of a combination of gonadotrophins and clomiphene are given. Ultrasound scanning is used to monitor the development of the follicles. When the follicles get to about 20 mm diameter it means that they are about to ovulate. An injection of hCC is given, and collection is timed for around 34 hours after this. Final oocyte maturation is critically dependent upon the timing of this final dose of hCG.

2 *Semen analysis and preparation*
On the morning of the woman's oocyte collection, the man produces a semen sample by masturbation into a sterile tube. He will already have had a full semen analysis, consisting of estimation of volume, sperm density, % normal forms, motility and presence of antisperm antibodies. The sperm to be used for the IVF are spun and washed, and counted. Approximately 100 000 motile sperm are required for the *in vitro* insemination.

3 *Oocyte collection*
There are two main methods for this: ultrasound guidance, which is the more usual, and laparoscopy. The ultrasound guidance can be done with the woman sedated, in which case she comes in for the oocyte collection as an out-patient. Light intensity in the theatre and embryology laboratory is kept to a minimum to protect the oocytes from ultraviolet light which is thought to be harmful to them. A fine needle is passed through the woman's bladder or vagina and guided by ultrasound to the ovary. A follicle is visualised, punctured, and the oocyte and its surrounding fluid gently sucked (**aspirated**) directly into a tube containing culture medium. This tube is immediately handed to the embryologist who examines the contents under a microscope and records any oocytes collected. The surgeon then proceeds to the next follicle, and the process is repeated until all the follicles on both ovaries are aspirated. After the oocyte collection is complete, the woman is allowed to rest until she is sufficiently recovered to go home.

Use of laparoscope (an instrument that allows the surgeon to look directly into the abdomen) for oocyte collection gives a very clear view of the follicles. A general anaesthetic is needed for this method and recovery is not so rapid. The collecting needle is inserted through the woman's abdominal wall separately, and the procedure for follicle aspiration is the same as before. The oocytes are placed in separate labelled tubes, then taken in a heated test tube block back to the laboratory.

4 *Insemination*
The oocytes are graded according to their maturity and each is put into a separate 50 microlitre droplet of fresh culture medium which is held at a steady pH and temperature in the incubator. The 100 000 sperm are added to each droplet 38–40 hours after the hCG injection. The sperm and oocytes are then returned to the incubator until the following day.

5 *Fertilisation and embryo culture*
The oocytes are inspected 12–18 hours after insemination to see whether they have fertilised. The appearance of two pronuclei indicates normal fertilisation. Re-insemination of apparently unfertilised oocytes sometimes brings about fertilisation, but any with one or three pronuclei are discarded as abnormal. Some treatment centres transfer the pronuclear embryos to the woman, but others return them to the incubator for a further 24 hours, during which time the first cleavage divisions occur.

Following this, the embryos are graded according to quality (figure 8.2). The best three are selected to be transferred back to the woman, and any others that are suitable may be frozen for possible replacement in subsequent cycles. These 'extra' embryos can also be donated to another couple who are unable to create their own, or alternatively given to research.

6 *Embryo transfer*
The woman lies comfortably on the bed, often with her partner nearby. As soon as the woman is ready, the embryos are drawn up into a fine flexible catheter with a small amount of medium. The embryos are gently flushed into the uterus via the cervix. The woman lies quietly for an hour or so, and then goes home. Embryo transfer is quite painless, and she is able to go about her usual life straightaway. Some treatment centres support the luteal phase of the cycle with additional progesterone in order to maximise the chance of implantation. Sensitive pregnancy tests can be given 14 days after embryo transfer. The chance of multiple pregnancy is quite high – 27% of all pregnancies by IVF in 1989.

As IVF is such a complicated procedure, reasons for failure can occur at any stage:

- the ovaries may fail to produce follicles (15%);
- oocyte collection may be impossible because of the inaccessibility of the follicles (5%);

Grade 1

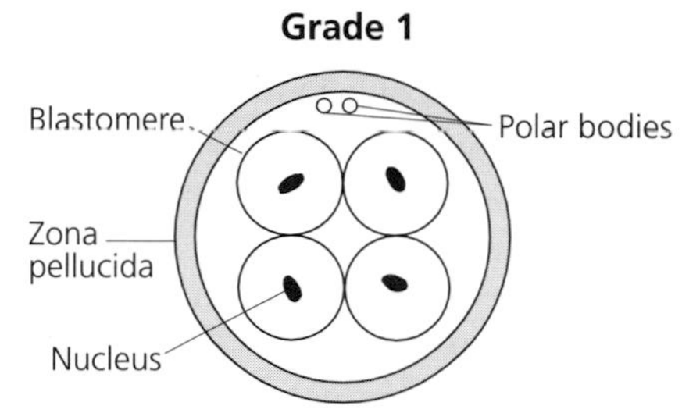

Embryo showing expected number of cells for time after insemination, individual blastomeres intact and symmetrical, with no fragmentation.

Grade 2

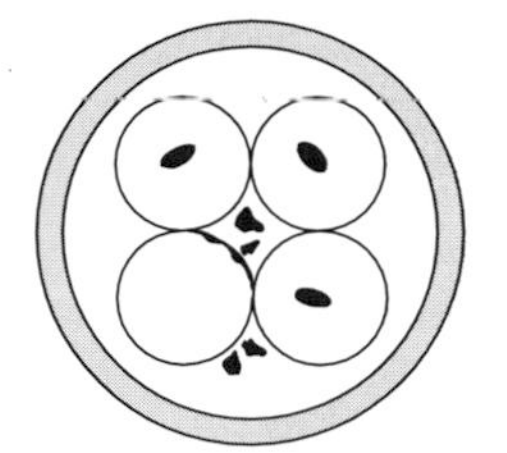

Embryo showing one of: slow development, asymmetrical blastomeres, extracellular debris or slight fragmentation.

Grade 3

Grade 4

Grade 5

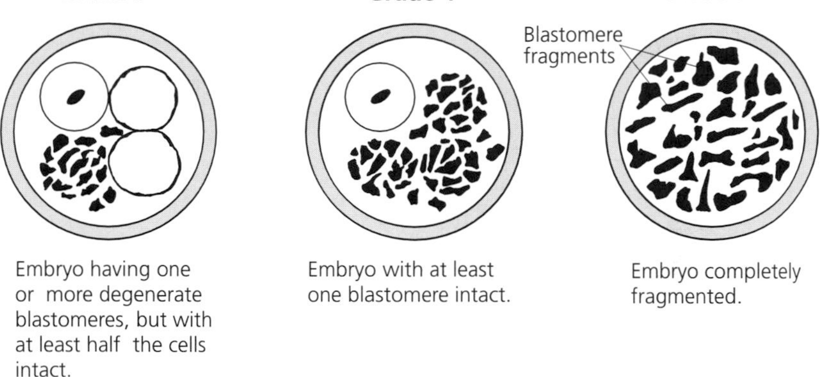

Embryo having one or more degenerate blastomeres, but with at least half the cells intact.

Embryo with at least one blastomere intact.

Embryo completely fragmented.

Figure 8.2 Embryo morphology grading system

- the oocytes may not fertilise (20–25%);
- the embryos may fail to develop normally (20%);
- the embryos may fail to implant, and a menstrual period follows. This is by far the greatest cause of failure of IVF, and the one which is least well explained. There is evidence that this happens relatively frequently in normally fertile women.

Some treatment centres believe in 'natural cycle IVF' in which no superovulatory drugs are employed. As the woman almost always produces only a single oocyte, her chance of success is significantly reduced.

GIFT

GIFT stands for gamete intra-Fallopian transfer, and involves oocyte collection and immediate replacement into the tube together with 2-300 000 sperm prepared in the same way as for IVF. It is less successful than IVF, but is a simpler and cheaper procedure, requiring less complex laboratory facilities.

Micromanipulation of oocytes and individual sperm

The outer coats of the oocyte present considerable obstacles to sperm which have poor motility or morphology. The surrounding cumulus cells can easily be dissected away by hand and the zona pellucida can be breached more easily to allow the entry of sperm. This can be done by forcing a very fine glass needle through it (partial zona dissection) (figure 8.3), or directing a let of acidic medium at it (zona drilling).

There are two additional procedures, both barely out of the experimental stage. They are extremely difficult technically, but have the potential to overcome virtually any fertility problem involving the gametes. **Sub-zonal insemination** (SUZI) involves introducing a single sperm cell between the zona pellucida and the oocyte surface membrane (figure 8.4a).

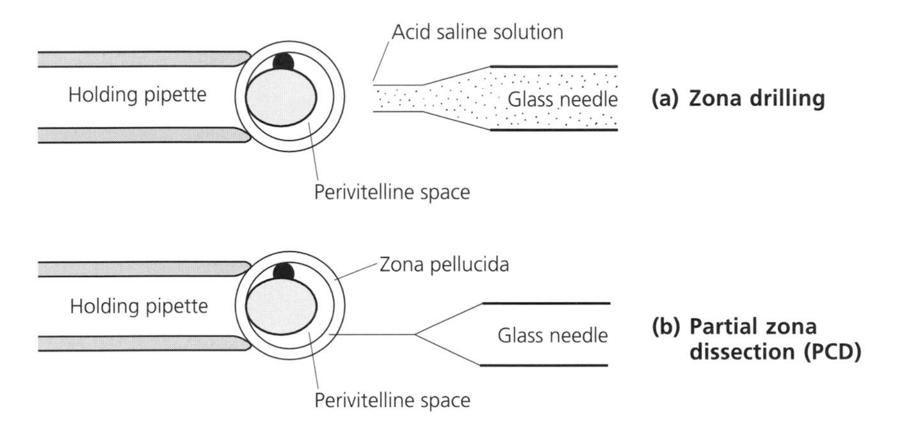

Figure 8.3 Methods for opening the zona pellucida to allow sperm to enter.

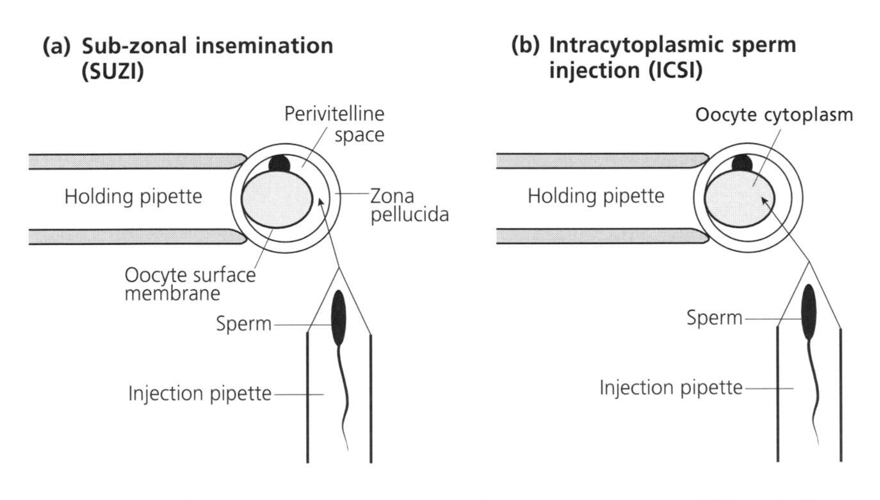

Figure 8.4 Techniques for micro-injection of sperm into oocytes. Although the success rate of these procedures is very low, pregnancies as a result of treatment using them have been reported.

The sperm still has to bind with the oocyte surface membrane to gain entry to the cytoplasm of the oocyte, but the barriers presented by the outer coats are overcome. As a method of assisted reproduction it is likely to be superseded by **intracytoplasmic sperm injection** (ICSI) in which mature, immotile or very immature sperm cells alike can be induced to fertilise an oocyte by being injected via a micropipette directly into the oocyte cytoplasm (figure 8.4b). The success rate of these methods is extremely low at the moment, but many people involved in the treatment of subfertility believe that they will be important in the future.

8.3 Problems of implantation

Implantation requires normal function of the endometrium. Endometriosis (see chapter 7) is one of the common causes of this type of subfertility. This condition can be treated either medically, in which case drugs are administered to suppress menstruation, or surgically by removing the affected areas of tissues. The former treatment usually takes at least 6–9 months, during which time conception cannot occur, and there is no guarantee that the endometriosis won't return as soon as normal hormonal function is restored. The outlook with the surgical approach is comparable. Mild endometriosis can be treated successfully, making pregnancy possible, but the chance of success decreases with increasing severity of the disease.

Some women with congential abnormalities of the uterus can have corrective surgery. Often, however, it is not clear why an embryo fails to implant though medical research, probing at the microscopic and even molecular level of the structure of the reproductive tract, indicates that abnormalities of the membranes of the endometrial cells may be responsible.

8.4 Sex selection

Techniques are being developed which may be able to separate sperm bearing an X chromosome from those bearing a Y chromosome. This means that, using assisted conception procedures such as IVF, it may become possible to choose the gender of the baby much more reliably. This has become known as sex selection. As yet this has not been achieved reproducibly with human sperm, in spite of the claims of some practitioners. Success has, however, been claimed with bull sperm, which have differently shaped heads compared with human sperm. This enables the bull sperm, when placed in an electromagnetic field, to be separated fairly reliably according to electrical charge, which appears to correspond to the chromosomal content of the nucleus. A discussion of the ethical implications of human sex selection can be found in chapter 9.

NINE

Ethical and moral considerations of fertility and infertility

9.1 Introduction

Four basic facts of life are:

1. organisms have no control over whether or not they become alive;
2. once alive they strive to say alive;
3. they have a strong and instinctive drive to reproduce (in humans this drive is fortified by a strong moral intuition);
4. they all eventually die.

The presumption in favour of life

Human beings, as organisms, are subject to these facts of life, but we are also able to rationalise and to moralise. We are able to recognise the basic instinct to preserve life in other people too, and appreciate that we have a moral duty to protect that instinct in them. This has been termed the 'presumption in favour of life', and for humans it is a fundamental fact of life: most people believe that matters relating to human life are of moral significance. As society has developed, this has become increasingly formalised in people's behaviour, to the extent that it is now enshrined in laws which are designed to protect life.

Reproduction is the single most distinguishing characteristic of life. It is the driving force of evolution, which is responsible for the diversity of plant and animal species on Earth. People are subjected to this force as much as any other organism, and a basic and powerful urge to reproduce is felt by most people at various times in their lives. In itself, however, this does not really answer the question 'why do people want to have children?' Of course, not everybody does want them: some have them by 'accident', whilst others decide consciously to avoid having a family. But for the majority of people, having children is a natural and essential part of their lives. It is not surprising, therefore, that all aspects of human reproduction have, over the course of history, been the subject of intense moral and ethical debate. As medical and scientific advances have enabled people to have more and more control over reproduction, and further means of controlling and promoting it have been discovered, the debate has intensified. Recently, however, rapid technical advances have resulted in intense moral discussion, and people are expressing their emotions, often in the form of outrage and anger, with increasing ardour and enthusiasm about the emerging issues.

Medical ethics

For scientists and clinicians working in reproductive medicine these moral and ethical questions have to be faced with each new advance that presents itself. Scientific research precedes new methods and treatments, and when human subjects are involved, medical scientists are guided by the recommendations of the Declaration of Helsinki (1964, revised 1975). The introduction of new medical procedures in the treatment of patients is strictly controlled by legislation. Hippocrates, a Greek physician of the fifth century BC, set out a code of ethics within which clinicians should work (the Hippocratic Oath). More recently, the Declaration of Geneva (1948, amended 1968) by the World Medical Association outlined the responsibilities of members of the medical profession in more contemporary terms. This has now superseded the Hippocratic Oath, and at the time of being admitted as members of the medical profession, new physicians pledge themselves to consecrate their lives to the service of humanity; practice their profession with conscience and dignity; give first consideration to the health of their patients; respect the secrets confided in them, even after their patients have died; not permit considerations of religion, race, party politics or social standing to interfere with their duty to their patients; and to maintain the utmost respect for human life from the time of conception.

9.2 Abortion

Beliefs about protection of human life

Discussion about abortion often centres around the same questions – at what stage does the duty to protect human life begin, and what does our duty to protect human life entail? When is it justifiable to interrupt the fulfilment of development? The teaching authority of the Roman Catholic Church currently holds that life begins at conception, and that a human life is distinguished from that of animals because humans have a soul. When, then, does the soul begin to exist? At differentiation of the primordial germ cells? At gametogenesis? At conception? At fertilisation, gastrulation, implantation, organogenesis, birth?

If conception is defined as the process of fertilisation followed by the establishment of pregnancy, then clearly this is not a single, instant event. In fact, the establishment of a separate genetic identity cannot really be said to have occurred until the time of syngamy – when the process of genetic fusion of the sperm and egg is complete. This occurs at about 30 hours after the onset of fertilisation. Some Christians believe that human uniqueness begins at 'quickening'. This is the time when the woman becomes aware of the movement of the fetus inside her. This awareness can be sudden, or the woman may realise slowly that the subtle 'flutterings' she can feel are not her own body movements. Quickening occurs, on average, at different times for boy and girl fetuses. Others believe that a human's separate identity is not achieved until birth is complete. In current law, the fetus has no rights until it is born. Most societies put the rights of a pregnant woman above those of a fetus, the argument being that the mother is a person and therefore has responsibilities whereas, legally, the fetus has no personality until it is born.

Abortion and the law

Therapeutic abortion is illegal in the Republic of Ireland, Portugal and some parts of South America. In countries where abortion is legal, the question is not so much whether it is right or wrong, but under what circumstances it is justified. The laws on abortion in the USA vary from state to state, but in general it is available on request up to the first three months of pregnancy (first trimester). In the second trimester, a pregnancy may be terminated on medical grounds only, and after that only if the mother's life is in danger. Orthodox Jewish law allows the killing of a fetus in order to save the mother until the birth of the head. In many Eastern European countries, abortion is viewed as a substitute for contraception. In the UK, the Infant Mortality Act and the 1967 Abortion Act made abortion legal up to 28 weeks from the first day of the last menstrual period:

- if the continuance of the pregnancy would put the woman's life at greater risk than termination;
- if the continuance of the pregnancy would put the woman's physical or mental health at greater risk than termination;
- if the continuance of the pregnancy risked the physical or mental health of any existing child(ren) of the pregnant woman;
- if there was a significant risk that if the fetus was born it would suffer from a serious physical or mental handicap.

This situation changed with the passing of the 1990 Abortion Act, which stipulates no upper limit for gestation in cases where there is a risk of fetal abnormality, but sets a limit of 24 weeks in cases where the woman's mental health may be affected. If a woman requests a termination of pregnancy, two doctors are needed to certify that one of the clauses of the

Act applies to her case. The father does not need to consent to the termination. The doctors also have to notify the Department of Health that they have carried out the procedure, and give information about the circumstances of the case. Over 100 000 legal abortions are carried out in the UK annually. It has been estimated that world wide, about 50 million abortions are performed per year, and more than 40% of these are by unsafe methods.

Babies born as early as 23 weeks of gestation can now survive, and in countries where abortion is legal this has called the 28 week limit into question. Some antenatal diagnostic tests, which may help the parents to decide whether or not to terminate a pregnancy, do not give reliable results until 18–20 weeks of gestation. Laws which only allow abortions before this time may not give the parents an opportunity to make an informed decision to terminate. However, the 28 week limit for abortion is largely determined not by the time at which the fetus becomes viable, but by the 'Mixed Strategy' view (see below) according to which the moral status of the fetus changes as it develops. Initially, the fetus is comparable, in most moral respects, with a piece of body tissue or an organ, whereas in later stages of development it is viewed morally equivalent to a newborn baby.

The arguments for and against

Some people with conservative views about abortion often do not discriminate between first and third trimester fetuses and wish to outlaw all abortion, even that carried out in the first weeks of pregnancy before all the neural structures allowing consciousness have developed. The essential belief of these people with strong anti-abortion views is that the fetus is, from the moment of conception, morally on a par with an adult human being, and that killing a fetus is therefore morally the same as killing an innocent person.

At the opposite extreme, some people with liberal, pro-abortion views argue that women should have the unquestioning right to terminate an undesired pregnancy at any stage. The essential belief of this group is that the fetus is, from conception until birth, morally equivalent to part of the woman's body, and that the woman has complete jurisdiction over it.

The Mixed Strategy adherents hold a view between the conservative and liberal standpoints on abortion. They accept that people have a duty to protect the embryo which increases with increasing developmental complexity, and that early abortion is less objectionable than late abortion. They have put forward the view that the present law could be adjusted such that far more stringent conditions are imposed as pregnancy proceeds. Further, if there is no diagnosis of fetal abnormality, there could be an earlier time limit for a requested abortion than exists under the present legislation.

Later therapeutic terminations of pregnancy may legally be carried out when there is risk to the life of the mother, but for those with conservative views on abortion this is morally on a par with killing an innocent person. Many holding the Mixed Strategy view also disagree with abortion on these grounds because they see it as comparable with hauling a hapless victim from the street and killing him or her to save the life of the mother (for example, by taking organs from the victim). The underlying assumption that the death of the mother is morally weightier than any other fate that might befall her in her continuing life has also been called into question. It is arguable that some life experiences are at least as important as death, and should therefore carry at least equal moral relevance. The philosopher John Stuart Mill (1806–73) argued that execution was a kinder form of punishment than some alternatives. Is it right to assume that human life has a value which is independent of its actual content?

Similarly, therapeutic abortion on the grounds of risk of injury to the physical and mental health of other children of the pregnant woman may also be morally unacceptable to the Mixed Strategy adherent, who would disagree with the liberal view that the fetus is morally on a par with a part of the body. This implies that having a late termination is like having, say, the appendix removed. It could be argued that these grounds for late termination constitute a social rather than a medical reason for abortion, and therefore that what is needed is social rather than medical intervention.

9.3 Contraception

The need for contraception

Many people accept that limiting the size of their family is not only necessary but desirable. Although no method is entirely without disadvantages (chapter 6), most couples, given sufficient understanding and with available contraception, come to some arrangement whereby they do not have unwanted pregnancies.

In many parts of the developing world, having very large families is not only part of the ordinary culture, but also a necessity in the face of a high mortality rate of babies and children. It has been estimated that the world population will increase in the next 30 years from 5 to 8 billion, with the developing countries showing the greatest rate of increase. This represents a huge world problem, and this was highlighted in 1992 by a joint declaration by the Royal Society in the UK and the National Academy of Sciences in the USA that the most pressing scientific priority in the next decade was the development of new methods of contraception.

Objections to contraception

The objection of the Roman Catholic Church to contraception is deeply rooted in its historical theology. In Genesis 38: 1–10 is the story of Onan, who was required by tradition to take on responsibility for his dead brother's wife and produce children to strengthen the family. He chose to engage in *coitus interruptus* rather than have children with his sister-in-law, because he wanted his own children to benefit. In the words of Genesis, God was displeased and killed Onan. The essence of Onan's crime was stealing his brother's inheritance as a result of 'spilling his seed'. This has been interpreted by the Roman Catholic authority as a strong argument in favour of not separating intercourse from procreation.

In the thirteenth century, St Thomas Aquinas put forward his idea that what God intends is right and proper. In other words, what is right or wrong for human beings is what God wants for us. In relation to reproduction, the interpretation of this by the Roman Catholic Church is that the natural course for a sperm is to fertilise, within the female receptacle.

Between 1930 and 1958 the Church of England agreed that its teaching on contraception should be that couples have a moral duty to limit the size of their families, and so the use of contraceptive methods was validated, and remains so to this day. This decision caused a division between the Churches of England and Rome. Successive popes, as heads of the Roman Catholic Church, published various encyclicals from 1930 onwards which stated their teaching on the subject of contraception. The latest encyclical, *Veritatis Splendor* published in 1993, affirms that contraception is forbidden by the Church as it separates copulation from procreation.

For many Roman Catholics this poses a great moral dilemma, because their tradition requires them not to act against their consciences. If, therefore, they believe that allowing repeated pregnancies and very large families is not what God intends for them (using St Thomas as their moral arbiter), then they may feel justified in using contraception, despite the teachings of the Church. Also, many devout Roman Catholics, whilst owning to the authority of the Pope, believe that they should not listen uncritically to his advice to them.

9.4 Assisted reproduction

Problems of fertility

Fertility is an intrinsic part of life, and infertility is viewed as a pathological state. Some infertile people have expressed the view that the fertile seem to take their inheritance for granted, and that the loss experienced by the involuntarily childless is a form of grief which is difficult to come to terms with because there is no focus for it – the grief is for the baby who never was. Infertility treatment is expensive, both financially and emotionally, invasive of the most intimate parts of people's lives and, for many, very unnatural, but such is the desperation of some childless people, they are willing to undergo many weeks, months and even years of treatment.

When a couple who decide to have a child discover that they are unable to conceive, for them it is a personal tragedy. If they are advised that assisted reproduction may offer them a chance of parenthood, they then have to decide whether it is physically, emotionally, morally and financially acceptable to them because, as has been pointed out, the procedures involved are not without considerable investment.

Those who are opposed to infertility treatment question the 'right' of everyone to have children. An argument often put forward is that expensive medical resources should not go into infertility treatment when there are many children already born who need parents. These resources, they contend, could be better expended in promoting improved parenting or in other forms of medical care. This argument is based on the assumption that bringing up children is more important than 'incubating' a baby. A more fundamentalist opposition to infertility treatment is based on the view that infertility helps to ameliorate the problem of world overpopulation. However, neither of these views take into account the consequences for the childless individuals whose genes will not be passed on to the next generation, a situation which has been called 'genetic death'.

Is it everyone's birthright to reproduce? The laws of natural selection are no respecter of people's emotional longings, and if there is a good biological reason for being unable to have children then, it might be argued, that is evolution's way of strengthening the species to ensure its survival. However, humans have overcome many problems of infertility with increasingly sophisticated and invasive techniques, and it could be argued that to help someone to reproduce who might otherwise be facing 'genetic death' is not only of considerable benefit to them, but also that any defective genes which would otherwise have died out will probably be dealt with easily by as yet unavailable scientific techniques.

Regulation of infertility treatment in the UK

Centres in the UK that offer assisted conception treatment are under licence issued by the Human Fertilisation and Embryology Authority. An important aspect of the code of practice to which the centre must adhere is that account should be taken of the welfare of the child which might result from the treatment. If the centre feels that for any reason the people seeking treatment are not in a position to bring up the child in reasonable circumstances, they are obliged under the law (Human Fertilisation and Embryology Act 1990) to withhold treatment. Of course, this is highly judgmental, and in this circumstance it is the assisted conception team, usually lead by a consultant gynaecologist, who decide whether that patient may be helped to conceive in that centre. Many centres, for example, refuse treatment to single or lesbian women on the grounds that the child has the right to a father. Other reasons for refusal of treatment may include health factors and criminal record. The issue of who gets treatment is highly controversial, not least because there are no clearly drawn guidelines. As part of their professional ethical code, doctors give precise and careful consideration to these matters, and are obliged as a profession to accept responsibility for their decisions – the buck stops with them.

Religious views on infertility treatment

Existing views about assisted reproduction based on religious beliefs centre around what is deemed to be natural/unnatural and whether that which is considered to be unnatural is therefore immoral. This is by no means the straightforward question it might appear to be. It could be argued that medical treatment, for example, is unnatural, but few would argue that it is immoral. The philosopher David Hume (1711–76) said: 'I am surely morally permitted to get out of the way of a falling rock which might otherwise kill me, despite the fact that the motion of the rock is the outcome of natural law'.

What is natural or unnatural with respect to human nature is even more difficult to disentangle, and this is mainly why religious disputes are among the most highly charged. It is part of the physical, psychological and emotional behaviour of humans to form pairs, reproduce by sexual intercourse and raise their own children – this is human nature. To try to outwit it by intervening in reproductive processes with, for example, abortion, contraception, assisted reproduction and masturbation could be said to be at best futile, and at worst dangerous and potentially destructive. However, in all species 'natural' implies 'normal', and not all individuals exhibit the norm. Common and familiar exceptions to 'the norm' exist: humans are sighted, but some people are born blind; humans are mammals, but not all women are able to breast feed; humans have a drive to reproduce, but some people are voluntarily childless; and some people kill their own children.

In the case of assisted reproduction the argument that it is unnatural bites both ways. If most humans have a natural desire to form reproductive pairs and raise their own young, then any attempt to thwart this desire is futile or even potentially calamitous, as is an attempt to do anything else that goes against human nature. Hence it can be argued that infertility treatment is acting in accordance with nature rather than against it.

For Roman Catholics, there is a moral steer on the topic of assisted reproduction from the Church. The view is that assisted reproduction is unnatural, because it separates the act of sexual intercourse from the act of procreation. Obviously, in order to produce a sperm sample for an assisted conception procedure involving gametes outside the body, masturbation is the most convenient method. Returning to the story of Onan, masturbation is seen to be sinful as well as *coitus interruptus* – both are versions of 'spilling the seed'. It is for this historic reason that masturbation is held to be a sin. Another reason is that, according to the Roman Catholic Church, the male genital organs are for procreation and urination only, therefore any other use of them (e.g. masturbation) is unnatural and thus unethical. Roman Catholics are not the only people with this view. In some African cultures masturbation is believed to compromise potency, and men refrain from it. This has implications for donor insemination, because men of this ethnic origin rarely volunteer to be sperm donors.

The use of donated gametes and tissues

Women with ovarian failure can be helped to conceive by the use of donated oocytes. Obviously, these are in much shorter supply than donated sperm, and scientists have responded to this problem by seeking new sources of eggs. At present, the only supply of donated oocytes is from women who are having fertility treatment themselves, women undergoing surgical sterilisation, or purely altruistic donors who volunteer for ovarian stimulation and egg collection. These donors are not legally allowed to receive payment, but are motivated purely by the desire to help the childless. Research suggests that ovarian grafts from live donors or from cadavers and ovarian tissue from aborted female fetuses could be used to treat some infertile women. Although it has now been decided not to allow it, public debate continues about whether such tissues should be used for either treatment or research. The main issues are:

- Should ways of increasing the number of eggs available for treatment or research be sought?
- Should tissues and eggs from cadavers and aborted fetuses be used in treatment or research?
- Who should give consent to use the tissues or eggs, when should consent be given, and in what form?

The Polkinghorne Report (1989) gave guidance on the research use of fetuses and fetal tissue. It was stated that tissue from a therapeutically (legally) aborted fetus could be used 'provided any decision concerning its use is separated from the decision to induce abortion'. It is the principle of separation which is important in this context, because it means that the abortion cannot be performed in order to provide tissues for other use, and the mother should not allow this reason to influence her decision to have an abortion. The report also suggested that a mother who had a miscarriage or a termination of pregnancy because the fetus had died should not be approached to give her consent for the use of fetal tissues.

For many people there is an intuitive revulsion towards the use of fetal tissue in this way. Some people are concerned about the cultural acceptance of a child conceived by donor insemination. What about the harmfulness of allowing a child to grow up with no knowledge of its origins? On the other hand, how would the child feel if it was to learn that its biological mother had been an aborted fetus? What about the possibility of the child rejecting its parents when it finds out the truth about its parentage? Many people believe that the law, based on moral principles, should ensure that no one should ever have to face this knowledge of their genetic heritage.

One of the most important issues surrounding assisted reproduction that involves the use of donated gametes – sperm or oocytes – is that of consideration for the welfare of the child born as a result of this kind of treatment. The HFEA will only grant a treatment licence to a centre which satisfactorily counsels its patients. Among the questions explored with the patients is whether, when and how to tell the child of the circumstances of his or her conception. Before DI became a relatively common procedure, it was very rare for parents to tell a child that he or she was conceived as a result of DI. However, since the HFE Act came into force in 1990 the legal status of the child has been protected, and patients are often advised that it is in the child's best interest to know about this aspect of his or her origin. When single women or lesbian couples request DI treatment, the centre is obliged to consider the matter of the child's need for a father before offering treatment. Many centres refuse to treat such women on these grounds.

In some rare cases gametes of a matching ethnic group are unavailable to a couple seeking treatment. This situation raises the same moral problems which occur when children of a different ethnic background are fostered or adopted by a family. Again, before offering treatment, the centre must give careful consideration to the circumstances surrounding any particular case of this kind. The HFEA's Code of Practice requires centres to take into account each couple's preference in relation to the general physical characteristics of the donor, which can be matched according to good clinical practice. It would, however, not be regarded as good clinical practice to treat a couple using gametes of a different ethnic origin from either the man or the woman.

It has recently become possible to treat post-menopausal women to enable them to have children. It has been argued that men can father children at a very advanced age, and so medical science should enable women to become mothers later in life as well. Some people see this as a great advantage for women who want to build their careers before having children. Others say that it is not fair on a child to have a much older that each case should be considered individually, and that it is not necessary or advisable to fix an upper age limit for the treatment of infertility.

The cost of infertility treatment

There are very few clinics in the UK at which patients can have *in vitro* fertilisation (IVF) on the National Health Service. This means that unless people live in certain health authority areas and can wait for a considerable period for an appointment, they must go to a private clinic if they want or need IVF treatment. This can be very expensive. There is no certainty, or even strong likelihood, of a positive outcome from IVF treatment. The success rates of even the best centres rarely exceed 20% live births per cycle of treatment (chapter 8). Most centres charge a fee per cycle, and a course of treatment can easily run into thousands of pounds. It could be argued on the one hand that if people who are otherwise healthy want this labour-intensive, 'hi-tech' treatment, then they should be prepared to pay for it, and on the other that everyone who wants a baby but is unfortunate enough to be unable to have one ought to be treated free. In fact, it is a measure of the desperation of the involuntarily childless that they are prepared to incur large debts in order to pay for the treatment. It is also a sad fact that the rate of marital breakdown of couples who have undergone IVF treatment is greater than the national average. Those couples who experience repeated failures of IVF treatment may experience an additional burden of grief over and above that caused by their childlessness.

The issue of who should pay for the cost of infertility treatment is based, at least in part, on the type of pathological condition it is considered to be. Is it:

1. an ordinary disorder like appendicitis which deserves NHS treatment;
2. a dysfunction of a particularly tragic and compelling kind, such as spina bifida;
3. a complaint of questionable status, such as baldness, which a person might want treated for relatively trivial reasons?

The question of what constitutes human nature, already discussed in this chapter, lends evidence to the argument for infertility falling into category 2. Most people seem to have a natural desire to have children, and intrinsic to this desire seems to be having children in the ordinary way, i.e. by sexual intercourse within a loving relationship, followed by pregnancy and labour of the genetic mother, then birth and nurturing of the children within the family. Infertile couples are unable to have children in this way, and are therefore unable to experience full satisfaction of this natural desire. By definition, no form of infertility treatment can satisfy a couple with all aspects of this desire: it is a damage-limiting approach.

Because it is possible to go on living and having a fulfilling relationship in marriage without children, this has lead to some people forming the opinion that infertility treatment comes into category 3 listed above. However, most couples who have received infertility treatment would say that it is better, sometimes very much better, than nothing.

Surrogacy

Some women who have functional ovaries but are unable to become pregnant have undergone egg retrieval and IVF with their husband's sperm, but made an arrangement with another woman to have the embryos transferred to her uterus, with the hope of pregnancy and a live birth. This is surrogacy, and a special provision has to be made in court to allow the genetic parents to become the legal parents, because in law the woman who bears and gives birth to a child is the legal mother, regardless of the child's genetic origin. The court will only grant the parental order if it is satisfied that no money has been involved in the surrogacy. In other words, the surrogate mother cannot be paid for carrying the other couple's child. This issue again brings forward the concept of 'having children'. It is the perception of some people that pregnancy, labour and birth are essential features of this process, but clearly, as in the case of surrogacy, it is possible for a child born naturally to one set of parents to be the genetic offspring of another couple. It is interesting to speculate from the ethical point of view exactly who has 'had' the baby after completion of the full procedure: drawing up the legal contract between the commissioning couple and the surrogate mother; gamete collection from the commissioning couple; the *in vitro* fertilisation by the embryologists and andrologists; the embryo transfer by the medical team; and the surrogate mother's pregnancy and birth.

9.5 Sex selection

There are two main ethical problems surrounding sex selection. The first concerns the medical reasons for doing it, and the second concerns the social reasons. In cases where a woman risks having a child with a life-threatening, sex-linked disease, choosing the sex of the child may ensure that it is born healthy. Most people, given a choice, would rather have a healthy child than one with a severe disability, and most people, given the same choice for themselves, would rather be healthy than disabled. However, this is not to diminish the value or contribution to society made by these disadvantaged people. On the other hand, in a family with, for example, one or more children of one sex, the couple may want to have another child of the opposite sex; alternatively, some cultures may attach higher status to one sex, and couples may wish to ensure that this is the gender of the child they have. Such social reasons for influencing the numbers of girls and boys born are not considered by the HFEA to be desirable.

The arguments put forward in favour of sex selection, namely:

- that there is no evidence that it may have an adverse effect on society;
- that couples wanting a child of a different sex may carry on having more children than they really want or can support;
- that sex selection is an important medical advance which it would be wrong to deny people who wanted it were felt to be outweighed by those against it;
- sex selection for social reasons is rarely in favour of girls. For example, inheritance of titles or wealth are often only legally possible in the UK through the male line, and in some ethnic populations there may be a considerable financial cost of having girls when it is time for them to marry;
- some people think that it is important for the first-born child in a family to be a boy. Position in the family is known to have an effect on a child's psychological development, and the effect of having a majority of male first-borns may reinforce adverse sexual stereotypes, perceptions of gender status and patterns of sex discrimination already present in society.

It will be seen readily that these are culturally based arguments against sex selection for social reasons. In the absence of this cultural backdrop, for example if society was matriarchal instead of patriarchal, or if social order was neither matriarchal nor patriarchal but equal, these arguments would not exist and there would be no need to deny sex selection. So the arguments against sex selection for social reasons are based on the admission that gender has value, a view considered not to be an appropriate attitude to encourage, and the practice of sex selection on these grounds is not permitted under the HFE Act. Now, sex selection is permitted for sound medical reasons, but only where secondary techniques, i.e. post-fertilisation (chapter 8), are used.

9.6 Embryo research

In the UK the Human Fertilisation and Embryology Act 1990 states that any research project involving the use of live human embryos must be licensed by the HFEA. Licences are only issued after thorough and careful scrutiny of the research proposal. The type of research which can be undertaken is strictly limited and very closely monitored by the HFEA as it proceeds. Licences are only given to projects if they are considered to:

- promote advances in the treatment of infertility;
- increase knowledge about the causes of congential disease and to develop methods for detecting genetically abnormal embryos before implantation;
- increase knowledge about the cause of miscarriage;
- develop more effective methods of contraception.

The whole debate about embryo research and whether or not it is morally acceptable really centres around the definition of an embryo, and this is related to the question of when life begins. In this respect, the arguments against research using embryos are similar to those against abortion. For those people who believe that a unique human life comes into being at the moment of conception, any procedure which interferes with the normal development of the embryo is unacceptable because they see the embryo as having the same rights and interests as any other child or adult.

The genetic identity of a new individual is established at syngamy, about 30 hours after the fusing of the sperm with the oocyte membrane, and it is at this stage that cleavage begins. It has been known for some years that human eggs can undergo cleavage in the absence of fertilisation, a condition known as **parthenogenesis**. In an experimental situation, development to the 25-somite stage has been observed from unfertilised mouse eggs. This calls into question the moral status of the embryo – is it based on genetic uniqueness? If so, then what about the parthenogenetically dividing embryo? Although this situation raises uncertainties over the moral status of the embryo, it is relevant to note that the human embryo undoubtedly has a genetically programmed uniqueness, giving it the potential for at least a reasonable life within humankind. It is in this context that the moral status of the embryo should be considered.

Experimental procedures on human embryos are only permitted under the law for the first 14 days offer the mixing of the gametes – in other words before the appearance of the primitive streak. This has led to the coining of the term 'pre-embryo'. Under the law, certain types of research on human embryos are prohibited. These include:

- replacing a human embryo in an animal;
- nucleus substitution – this procedure comprises removal of the nucleus of an embryonic cell (blastomere) and replacing it with a nucleus taken from the cell of another person or embryo;
- altering the genetic structure of any cell while it forms part of an embryo;
- cloning human embryos for the purposes of infertility treatment.

Other than in these circumstances, it is only after this period of 14 days from fertilisation that human life is legally protected. The embryo which has been experimented upon cannot be maintained, either *in vitro*, frozen, or replaced into a woman, after that time, and it must be destroyed.

The mixing of human gametes with those of another species is also prohibited unless a licence is granted. A diagnostic test known as the hamster egg penetration (HEP) assay is sometimes used to see how well sperm can penetrate an egg. The hamster oocyte is unique in that when its zona pellucida is removed chemically it will permit binding and penetration of a wide variety of sperm from other mammalian and even non-mammalian species. In the HEP assay, 40 hamster eggs are incubated with the test sperm and after 3–4 hours the eggs are fixed and stained, and the percentage of eggs with decondensing sperm heads in their ooplasm is calculated. Although this procedure gives some measure of the fertilising ability of the sperm, it is by no means routinely done in centres offering assisted reproduction treatment. However, it is a useful research tool under certain circumstances, but requires a licence under the Act.

Such is the sensitivity of human embryo research, that the development of any new technique (chapter 4) must be strictly regulated. Research licences are only granted when it can be demonstrated that the researchers will add to the body of knowledge, and in doing so gain technical competence. Only then can they be allowed to use a new technique in the treatment of patients. It is with this careful and painstaking approach, coupled with on-going public debate, that big issues, such as what has been termed by the media 'designer babies', can be approached responsibly. There is no doubt that research will continue to strive to improve on nature; indeed, the elimination of 'undesirable' genes may be one of the consequences of the Human Genome Project, a world wide scientific investigation which aims to identify and map all the 100 000 genes on the human chromosomes by the year 2005. However, the quest to eliminate so-called 'undesirable' genes should always be tempered with the question, 'what is undesirable?'

A description of eugenic practices aimed at human improvement by genetic means is given in *Heredity and human diversity* by Stephen Tomkins (CUP, 1993) in this series. Here the eugenic and dysgenic (i.e. active selection of unfavourable characteristics) arguments are set out. Positive eugenic practice aims at selecting desirable phenotypes, whereas negative practice aims at avoiding undesirable ones. Although there is little doubt that the elimination of certain genes, such as those for Duchenne muscular dystrophy, cystic fibrosis or haemophilia, from the human population is desirable, it does raise the spectre of genetic and ethnic 'cleansing'. Lessons learned from past and recent history must inform each step in the implementation of advances in reproductive technology.

9.7 Conclusion

Rapid advances in reproductive technology and medicine have outstripped moral debate on the issues raised, with the result that almost weekly there is a new media sensation in this area of scientific research. However, this is not to say that people are slow to come to terms with the importance of new techniques or possibilities for treatment. The ethics surrounding the issues of human reproduction are at the same time fundamental and extremely complex; this makes them difficult, if not impossible, to be judged completely right or wrong. The subsequent controversies give rise to real dilemmas because of technical difficulties, uncertain outcomes and variance of principles.

Critical reading and frequent discussion help to inform a person's views. This is very important because although laws exist on such matters as abortion and embryo research, legislation should always reflect the views of society as a whole, and changes in the law can be made to reflect shifts in people's opinions about what is acceptable. This is not to say that the relationship between social morality and the legislature is static: indeed, one informs the other. Engaging in debate and voicing opinion is particularly important when new guidelines are being drawn up. The HFEA regularly puts out public consultation documents on such issues as sex selection and the use of donated ovarian tissue in embryo research and assisted reproduction, and the questions raised by these matters are discussed by treatment centres, church groups, political gatherings, academic departments and so on. Above all, it is most important for everyone to keep trying to come to terms with these difficult moral problems, by thinking carefully about their significance not just for oneself and for society as a whole, but also with empathy for those more directly and personally affected.

Index